USING THE PROJECT MANAGEMENT MATURITY MODEL

USING THE PROJECT MANAGEMENT MATURITY MODEL

Strategic Planning for Project Management

SECOND EDITION

HAROLD KERZNER, Ph. D.

WILEY

John Wiley & Sons, Inc.

Copyright © 2005 by John Wiley & Sons. All rights reserved.

Published by John Wiley & Sons, Inc., Hoboken, New Jersey
Published simultaneously in Canada

For general information on our other products and services or for technical support, please contact our Customer Care Department within the United States at (800) 762-2974, outside the United States at (317) 572-3993 or fax (317) 572-4002.

Wiley also publishes its books in a variety of electronic formats. Some content that appears in print may not be available in electronic books. For more information about Wiley products, visit our web site at www.wiley.com.

Library of Congress Cataloging-in-Publication Data

Kerzner, Harold.
    Using the project management maturity model : strategic planning for project management / Harold Kerzner.—2nd ed.
       p. cm.
    Rev. ed. of: Strategic planning for project management using a project management maturity model. c2001.
    Includes bibliographical references and index.
    ISBN 0-471-69161-5 (cloth)
     1. Project management. 2. Strategic planning. I. Kerzner, Harold. Strategic planning for project management using a project management maturity model. II. Title.

HD69.P75K494 2005
658.4'04—dc22   2004060627

Printed in the United States of America.

10 9 8 7 6 5 4 3 2 1

Contents _____

14 How to Conduct a Project Management Maturity Assessment 223

15 Understanding Best Practices 237

Case Studies 247

Foreword

Strategic workforce planning ensures that the capacity and capability of an organization's workforce is in place to accomplish the organization's mission, goals, and objectives. It builds a strong performance culture with measurable and sustainable growth.

As the business owner for the project management profession, the charter of this office is to manage and measure the *intangible assets* of the profession in order to meet *strategy execution* and accordingly, *create value* in the customer partner experience. Intangible assets mean building on human capital such as the job family of the certified project manager and supporting the performance-based approach to building sustainable competence. Strategy execution means providing accountability, communication, and teamwork aligned to our mission. Moreover, it creates value through behavior that drives toward results wherein all stakeholders (the customer, employee, partner, and vendor) attain success.

Project management is a shared competency, blended into many job titles and job clusters. At the same time it is, uniquely, a credentialed profession. Sales executives use it to articulate the goods and services sold. Consulting managers use it to construct a statement of work. Consultants use it to manage their time and communicate progress on their deliverables. Project managers use it to orchestrate delivery within the constraints of scope, schedule, and resources. Executives use it to understand their rhythm-of-the-business to measure and value risk and fiscal performance. The world has always lived with the contribution of project management. Organizations and governments realize, now more than ever, that they cannot survive the twenty-first century without formally addressing it as core to their workforce. Sustainable competitive advantage, Dr. Kerzner points out, is cross-functional within the organization, and this positions project management as a shared and core competency.

Prior to his introduction of the Project Management Maturity Model (PMMM), Dr. Kerzner provides a foundation for strategic workforce planning with a historical viewpoint reflecting on industry, the factors that influence and impact organizations as they deliberate project management, and practical principles for positioning and applying project management competency within an organization. This has been valuable for our own understanding and relevance: these chapters are a preamble that provides a compare-and-contrast view into project management in terms that our shareholders and stakeholders can understand.

Ensuing chapters that address the PMMM are clear and structured. In addition, the close of each chapter provides an informative assessment instrument that may be used and analyzed to assess where an organization stands, to serve as a discussion point for interpretation of the assessed results. The research in these chapters alone provides organizations with a tool to understanding the capability of their workforce and, importantly, their corresponding maturity.

In November 2002, Microsoft participated in the Kerzner Project Management Maturity Assessment for all our geographical regions. The results provided a favorable understanding of how we rated in the industry and with our own internal concentration. We found that a common language, the value proposition for project management, and its taxonomy, is an essential currency for all organizations worldwide. That common process reinforces predictability and quality, and as Bill Gates once said, "Every new project should directly build on the learning from any similar project undertaken anywhere else in the world." A thorough intellectual capital exchange for all aspects of intellectual property and information gained from project reviews is assurance to improved cycle time in process and in demonstrated quality of project deliverables. This was a significant contribution of maturity based on Dr. Kerzner's work and the result of three years of effort at Microsoft.

Our current challenge is in implementing a singular methodology and benchmarking, concurrently, not only for Microsoft, but also to positively influence our customer partner experience. Two important public domain programs built at this level of maturity are the Project Management Assessment and Credentialing (PMAC) Program and the Role Education and Mentoring Program, both referencing the activities and advancement criteria from Dr. Kerzner's work. We are confident this level of maturity will be a cornerstone to building human capital aligned to our mission—to enable people and businesses throughout the world to realize their full potential.

The message is clear: behavioral excellence and organizational maturity are constituent in the guidance Dr. Kerzner has provided in this work.

Christian A. Jensen, PMP
Microsoft, Business Owner, Project Management Profession

Foreword to the First Edition ⸺

To win a decathlon requires the extreme best from the participant. It is a very grueling and demanding set of events. The decathlete is usually very good and in fact the best in one or two events and in good standing in the other eight or nine events. The objective is to be the overall best in all ten events. Decathletes, like most athletes, must complete in head-to-head events to know if they are able to win the overall decathlon. They must study their competitors in the greatest detail and know their strengths and weakness. They must learn from the other decathletes what allows them to put out that extra 5 percent that means the difference between winning and just participating. They must also compete in an environment where the performance standard required to win is always becoming higher.

Being a project manager is similar to being a decathlete, and in the business of projects, the field is very competitive. Similar to a decathlon there are events (nine knowledge areas) in the Project Management Body of Knowledge. The decathletes in project management are the companies that are controlling costs, schedule, and quality on a project level. The project-driven companies must find ways to learn "best practices" in a competitive world and apply these lessons to their processes, systems, and tools. This method of continuous improvement through measuring and comparing is referred to as *benchmarking* as described by Dr. Kerzner.

Nortel aspires to win the decathlon prize, but realizes it is not possible without both internal and external benchmarking measurements and continuous improvements. The internal benchmarking is similar to intramural decathlons where learning comes from watching the friendly decathletes. A significant opportunity for learning and continuous improvement occurs when the "best in class" have entered the decathlon.

Nortel has participated in the Kerzner five-step Project Management Maturity Model survey for the last year and a half. The five steps measure the desired direction of Nortel in reaching the point of external benchmarking and continuous improvement. Over 400 Nortel Project Managers globally have participated in step one. Step one determines if a common language is being used. Nortel is using this initial assessment as a baseline for improvements. In the fall of 2000 Nortel will confirm the improvements in Level 1 and move to measure Level 2, "common process," and Level 3 "common methodology."

Dr. Kerzner has provided the measuring devices for the project management decathlon in the Maturity Model. By placing the sense of urgency around the improvement programs and remeasuring against the initial baseline, Nortel has a tangible measure of improvement and is encouraged to go on and participate in the external benchmarking in order to become the best in Project Management.

Dr. Kerzner's Project Management Maturity Model is on the internal web in Nortel, and the results are automatically calculated to provide the participant with an immediate score. The aggregate scores of each business unit are published monthly as a visible proof that Nortel is focused on the ideals of the five-step model (common language, process, methodology, benchmarking, and continuous improvement). The model leads to a strong foundation for a world-class, project-driven company to mature and evolve.

It has often been said that "to improve, one must be prepared to measure the improvement" and "one must inspect what one expects." The Kerzner Project Management Maturity Model has provided this tangible measure of maturity. The rest is up to the company to set the expectations and to inspect the results.

Bill Marshall
Formerly with Nortel Global Project Process Standards

Preface ————————————————————————————————————

Excellence in project management cannot occur, at least not within a reasonable time frame, without some form of strategic planning for project management. Although the principles of strategic planning have been known for several decades, an understanding of their applicability to project management is relatively new. Today, as more companies recognize the benefits that project management can provide to their "bottom line," the need for strategic planning for project management has been identified as a high priority.

This book is broken down into two major parts. The first part, Chapters 1 to 3, discusses the principles of strategic planning and how it relates to project management. The second part, Chapters 4 to 10, details the project management maturity model (PMMM), which will provide organizations with general guidance on how to perform strategic planning for project management. The various levels, or stages of development, for achieving project management maturity, and the accompanying assessment instruments, can be used to validate how far along the maturity curve the organization has progressed. The PMMM has been industry validated. One large company requires that, each month, managers and executives take the assessment instrument exams and then verify that progress toward maturity is taking place from reporting period to reporting period. Other companies have used PMMM to assess the corporation's knowledge level on project management as well as a means for assessing the needs for a project management office, a best practices library, external and internal benchmarking, a way of incorporating Six Sigma into project management activities, and the identification of the type of project management training needed.

Chapters 12–15 discuss some relatively new concepts in project management. Many of these concepts are the result of strategic planning for project management activities.

Perhaps the major benefit of the PMMM is that the assessment instruments for each level of maturity can be customized for individual companies. This customization opportunity makes *Using the Project Management Maturity Model* highly desirable as a required or reference text for college and university courses that require the students to perform an individual or group research project. The book should also be useful as a required text for graduate courses on research methods in project management. In addition, the book can be used as an introduction to research methods for project management benchmarking and continuous improvement, as well as providing a brief overview of how to design a project management methodology.

Seminars on strategic planning for project management using this book, as well as other training programs on various project management subjects, are available by contacting Lori Milhaven, Vice President, at the International Institute for Learning, 212-515-5121. Contact can also be made through the Web site (iil.com).

Harold Kerzner
International Institute for Learning
110 East 59th Street
New York, NY 10022-1380

Introduction _____

Using the Project Management Maturity Model:
Strategic Planning for Project Management, Second Edition

Prior to joining the International Institute for Learning, Inc. (IIL), I worked with a global consulting organization where senior management recognized the importance of project management only after one of our key clients requested that our solutions be delivered more quickly and at a lower cost. At the same time, our consulting group was requested to maintain or improve the overall quality of our solutions for our customers. In hindsight, perhaps we should have been more proactive and not have waited for our client to request improved project management practices and performance. However, it did happen and—fortunately in the end—we revolutionized the way the organization operated and interacted with our client base. As a result, the consulting group was then better positioned in the marketplace due to the exponential increases in our efficiency and effectiveness, which made our clients even more loyal to us than ever.

One of the first activities our Delivery Management leadership team performed was an objective assessment of project management maturity in our organization, as described in Dr. Kerzner's first edition of this book, then titled *Strategic Planning for Project Management Using a Project Management Maturity Model.* The result verified the theory that our organization was challenged in project management and determined that we were barely at Level 1 maturity.

A closer look at the organization uncovered different approaches to managing projects, depending on who was in charge: account executives, project coordinators, production managers, database administrators, or project managers. In addition to varied approaches in managing projects, further examination revealed that we were not communicating at all, due to great differences in our business language! Why was this happening?

Well, for starters, all of our employees, regardless of their functional group, were involved in project-related work one way or another, acting as project managers or project performers. Indeed, by the very definition of the term "project," my previous company, as well as other companies worldwide, employs project managers and project performers to create or modify products, improve business processes, enrich service delivery, launch new initiatives, build capacity, and so on. This means that employees at the beginnings of their career, those who were advancing from a technical to a generalist role, those who were reaching the pinnacle at the company, and everyone in between, work on mission-critical work for our clients. But how can we be successful if we cannot even understand each other?

Having a common language is where organizations must start on the road to project management maturity. Without a doubt, Dr. Kerzner's staged model provides a foundation for excellence for companies today where everyone is involved in project-related work. He presents to project management practitioners a practical PMBOK® Guide–aligned standard that identifies the organization's characteristics at each level in the maturity model, details what it takes to move from one level to another, and shares first-rate insights into how project management maturity capability is a component in every business' strategic initiatives.

Now, as the Director of Methodology and Assessment Solutions at IIL, I am pleased to spread the gospel of success achieved by my previous company, and by other world-class companies, who used Dr. Kerzner's project management maturity model. With this book as a starting point, IIL's Kerzner Project Management Maturity Model (KPMMM) Online Assessment tool provides the individual assessment participants and their organizations with a breakdown on how they are doing in different categories in each level, a comparison of their overall results against those of other companies and individuals who have taken the assessment, and a high-level prescriptive action plan to follow for individual and organizational improvement. IIL's web-based resource objectively identifies your strengths and weaknesses in project management, shows areas of organizational improvement opportunities, provides instant feedback and recommendations, and can be taken at your project managers' convenience.

IIL shares Dr. Kerzner's enthusiasm and vision to help organizations achieve higher levels of organizational project management maturity. As such, we hope that you will visit IIL's website to view the demo of this invaluable companion tool (http://www.iil.com/Assessment_Tool_Demo/welcome.htm). Or if you wish, you can contact us at assessment@iil.com or (212) 758-0177 to find out how we can be of service to you in your goal of achieving organizational project management excellence.

Vladimir Manuel, PMP
Director
Methodology and Assessment Solutions
International Institute for Learning, Inc.

The Need for Strategic Planning for Project Management

INTRODUCTION

For more than 40 years, American companies have been using the principles of project management to get work accomplished. Yet, for more than 30 of these years, very few attempts were made to recognize project management as a core competency for the company. There were three reasons for this resistance to project management. First, project management was viewed as simply a scheduling tool for the workers. Second, since this scheduling tool was thought to belong at the worker level, executives saw no reason to look more closely at project management, and thus failed to recognize the true benefits it could bring. Third, executives were fearful that project management, if viewed as a core competency, would require them to decentralize authority, to delegate decision-making to the project managers, and thus to diminish the executives' power and authority base.

MISCONCEPTIONS

As the 1990s approached, project management began to mature in virtually all types of organizations, including those firms that were project-driven, those that were non–project-driven, and hybrids. Knowledge concerning the benefits project management offered now permeated all levels of management. Project management came to be recognized as a process that would increase shareholder value.

This new knowledge on the benefits of project management allowed us to dispel the illusions and misconceptions that we had believed in for over 30 years. These misconceptions or past views are detailed below, together with current views.

Cost of Project Management

- Misconception: Project management will require more people and increase our overhead costs.
- Present view: Project management allows us to lower our cost of operations by accomplishing more work in less time and with fewer resources without any sacrifice in quality.

Profitability

- Misconception: Profitability may decrease.
- Present view: Profitability will increase.

Scope Changes

- Misconception: Project management will increase the number of scope changes on projects, perhaps due to the project manager's desire for creativity.
- Present view: Project management provides us with better control of scope changes. Good project managers try to avoid scope changes.

Organizational Performance

- Misconception: Because of multiple-boss reporting, project management will create organizational instability and increase the potential for conflicts.
- Present view: Project management makes the organization more efficient and effective through better organizational behavior principles.

Customer Contact

- Misconception: Project management is really "eyewash" for the customer's benefit.
- Present view: Project management allows us to develop a closer working relationship with our customers.

Problems

- Misconception: Project management will end up creating more problems than usual.
- Present view: Project management provides us with a structured process for effectively solving problems.

Applicability

- Misconception: Project management is applicable only to large, long-term projects such as in aerospace, defense, and construction.

- Present view: Virtually all projects in all industries can benefit from the principles of project management.

Quality
- Misconception: Project management will increase the potential for quality problems.
- Present view: Project management will increase the quality of our products and services.

Power/Authority
- Misconception: Multiple-boss reporting will increase power and authority problems.
- Present view: Project management will reduce the majority of the power/authority problems.

Focus
- Misconception: Project management focuses on suboptimization by looking at the project only.
- Present view: Project management allows us to make better decisions for the best interest of the company.

End Result
- Misconception: Project management delivers products to a customer.
- Present view: Project management delivers solutions to a customer.

Competitiveness
- Misconception: The cost of project management may make us noncompetitive.
- Present view: Project management will increase our business (and even enhance our reputation).

WALL STREET BENEFITS

The benefits recognized by the present views of project management are now seen to be strategic initiatives designed to enhance shareholder value. Perhaps one of the best examples showing this is the effect on stock price illustrated in Figure 1–1. An executive who wishes to remain anonymous believes that the difference between the target selling price of his company's stock and the actual selling price can be attributed to the quality of the company's project management system and management's ability to execute projects within time, cost, and quality constraints and to the customer's satisfaction. If the actual selling price was below the target selling price, it might indicate that the company, especially if it were

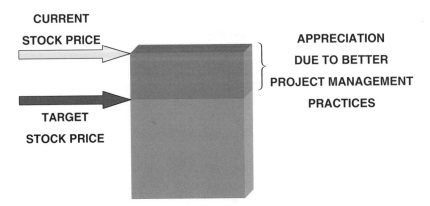

FIGURE 1–1. Impact on stock price as a result of better project management.

project-driven, was having fundamental problems with project execution, which would affect competitiveness and profitability.

The concept behind Figure 1–1 may seem plausible from a theoretical point of view. In reality, other forces may exist that can have a significant impact on the stock price, such as recessions, lack of new products, competitor's activities, legal problems, and ratings by financial institutions.

It may take years for a company just beginning to adopt project management to reap the potential benefits shown in Figure 1–1. Some of the organizations that believe they are achieving the benefits of Figure 1–1 are in these fields:

- Automotive subcontractors, some of whom are now treated as "partners" by their customers due to the quality of their project management systems.
- Financial institutions, especially those that are aggressively acquiring and assimilating other organizations and rapidly integrating both cultures into one without any appreciable negative effect on earnings.
- High technology companies who have beaten their competitors to the marketplace with new products.

Not all companies have the ability to reap the benefits of project management. Some do not yet recognize the benefits of or need for strategic planning for project management. Others recognize its importance but simply lack expertise in how to do it. In either event, strategic planning for project management is a necessity.

STAKEHOLDERS

Given the fact that project management is no longer seen as just a quantitative tool for the employees, but is recognized as a source of benefits to the whole corporation, project management must satisfy the needs of its stakeholders. Stakeholders are individuals or groups that either directly or indirectly are affected by the performance of the organization. These individuals are not only affected by the organization's performance, but may even have a claim on its performance. As an example, unions can have a strong influence on how a project management methodology is executed. The general public and government agencies may be affected through health, safety, and ethical issues in the way projects are executed.

Although there are several ways to classify stakeholders, the most common method is as follows:

Financial Stakeholders
- Stockholders
- Financial institutions (suppliers of capital)
- Creditors

The Product/Market Stakeholders
- Primary customers
- Primary suppliers
- Competitors
- Unions
- Government agencies
- Local government committees

Organizational Stakeholders
- Executive officers
- Board of Directors
- Employees in general
- Managers

Any strategic planning efforts must focus on the best interests of all of an organization's stakeholders, not merely a few.

GAP ANALYSIS

There are two primary reasons for wanting to perform strategic planning for project management. First and foremost is the desire to secure a competitive advantage. The second reason is to minimize the competition's competitive advantage or to strengthen your own competitive advantage.

The key to reducing any disadvantage that may exist between you and your competitors is the process known as gap analysis. Figure 1–2 illustrates the basic concept behind gap analysis. You can compare your firm either to the industry average or to another company. Both comparisons are shown in Figure 1–2.

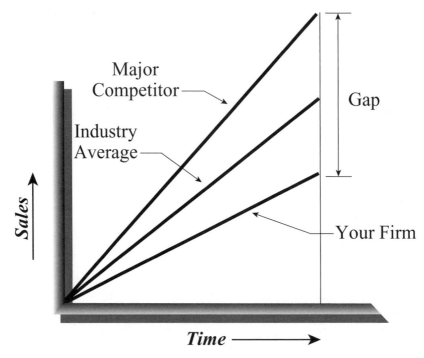

FIGURE 1–2. Gap analysis.

Just for an example, using Figure 1–2, we can compare the gaps in total sales. According to Figure 1–2, the gap between your firm and your major competitor is significant and appears to be increasing. The gap between your organization and the industry average is also increasing, but not as greatly as the gap between you and your major competitor.

For a company aspiring to perform strategic planning for project management, there are three critical gaps to analyze:

- Speed to market
- Competitiveness on cost
- Competitiveness on quality

Figure 1–3 shows the gap on speed to market or new product development times. If the gap is large between you and either the industry average or your major competitor, then to win the battle you must develop a project management methodology that allows for the overlapping of life cycle phases combined with appreciable risk-taking. The larger the gap, the greater the risks to be taken. If the gap cannot be closed, then your organization must decide if its future should rest on the shoulders of a "first-to-market" approach or if a less critical "me-too" product approach is best. Another unfavorable result would be the firm's inability to compete on full product lines. The latter could impact the firm's revenue stream.

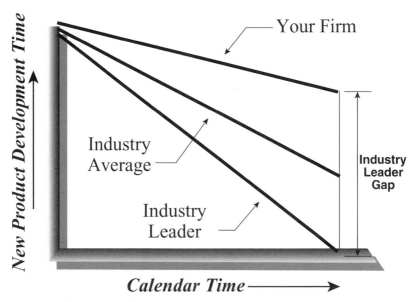

FIGURE 1–3. Gap analysis (time).

Another critical aspect of the schedule gap analysis shown in Figure 1–3 is customer's future expectations. Consider, for example, the auto manufacturers and their tier one suppliers. Today, these organizations operate on a three-year life cycle from concept to first production run. If you were a tier one supplier, however, and you found out that your primary customers were experimenting with a 24-month car, then you would need to perform strategic planning, not only to be competitive but also to be able to react quickly should your customers mandate schedule compression.

A gap on cost is an even more serious situation. Figure 1–4 illustrates the cost or pricing gap. Strategic planning for project management can include for provisions in the methodology for better estimating techniques, the creation of lessons learned files on previous costing, and possibly the purchasing of historical databases for cost estimating.

Good project management methodologies allow work to be accomplished in less time, at lower cost, with fewer resources, and without any sacrifice in quality. But if a cost/pricing gap still persists despite good project management, then the organization may either have to be more selective about which projects it accepts or choose to compete on quality rather than on cost. The latter assumes that your customers would be willing to pay a higher price for added quality or added value features.

Gaps on time and cost may not necessarily limit the markets in which you compete. However, gaps on quality, as shown in Figure 1–5, can severely hinder

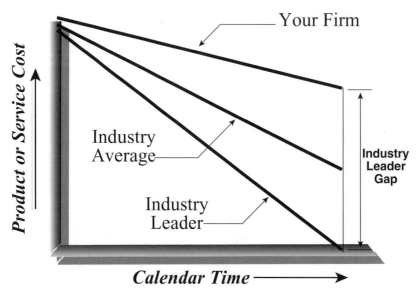

FIGURE 1–4. Gap analysis (cost).

your firm's ability to compete. The critical gap in Figure 1–5 is the difference between the customer's expectations of quality and what you can deliver. Good
project management methodologies can include policies, procedures, and guidelines for improving quality. However, the gap on quality takes a lot longer to compress than the gaps on time and cost.

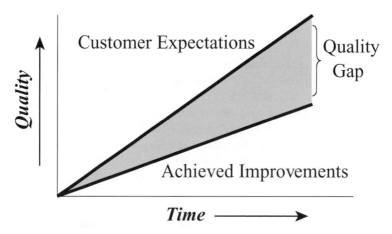

FIGURE 1–5. Gap analysis (quality).

CONCLUDING REMARKS

Strategic planning for project management, combined with a good project management methodology, can compress the gaps on time, cost, and quality. However, there are still critical decisions that must be made. Marketing must decide what products to offer and which markets to serve. The information systems people must assist in the design, development, and/or selection of support systems. And senior management must provide sufficient and qualified resources.

Strategic planning for excellence in project management needs to consider all aspects of the company: from the working relationships among employees and managers and between staff and management, to the roles of the various players (especially the role of executive project sponsors), to the company's corporate structure and culture. Other aspects of project management must also be planned. Strategic planning is vital for every company's health. Effective strategic planning can mean the difference between long-term success and failure. Even career planning for individual project managers ultimately plays a part in a company's excellence, or its mediocrity, in project management. All of these subjects are discussed in the following chapters.

Impact of Economic Conditions on Project Management

INTRODUCTION

Economic conditions can be favorable or unfavorable. Yet in either case, an astute company can convert someone else's misfortune into its own good fortune. Every place we look we find windows of opportunity. But to take full and prompt advantage of these windows of opportunity, to be truly successful, management must have a repeatable process predicated upon speed and quality of execution. The problem with most companies is that setting strategic targets can occur quickly, but developing implementation plans and executing them are much slower processes. Why did it take us so long to truly recognize the benefits of project management?

HISTORICAL BASIS

During favorable economic times, changes in management style and corporate culture occur very slowly. Executives are reluctant to "rock the boat." But favorable economic conditions don't last forever. The period between recognizing the need for change and garnering the ability to manage change is usually measured in years. As economic conditions deteriorate, change occurs more and more quickly in business organizations, but still not fast enough to keep up with the economy. To make matters worse, windows of opportunity are missed because no project management methodology is in place.

TABLE 2–1.   EFFECTS OF THE 1989–1993 RECESSON ON THE IMPLEMENTATION OF PROJECT MANAGEMENT

Factor	Prior to the Recession	After the Recession
Strategic focus	Short-term	Long-term
Organizational structuring	To secure power, authority, and control	To get closer to customers
Management focus	To manage people	To manage work and deliverables
Sponsorship	Lip service sponsorship	Active
Training emphasis	Quantitative	Qualitative/behavioral
Risk analysis	Minimal effort	Concerted effort
Authority	In writing	Implied
Team building	Functional teams	Cross-functional teams

Before the recession of 1989–1993, U.S. companies were willing to accept the implementation of project management at a tedious pace. The implementation, if it happened at all, simply consisted of using or adopting new planning and scheduling tools for the benefit of the employee, not the company. Corporate managers in general believed that their guidance was sufficient to keep their companies healthy, and outside consultants were brought in primarily to train production workers in the principles of project management. Executive training sessions, even very short ones, were rarely offered.

During the recession, senior managers came to realize that their knowledge of project management was not as comprehensive as they had once believed. Table 2–1 shows how the recession affected the development of project management systems.

By the end of the recession, in 1993, many companies had finally recognized the importance of both strategic planning and project management, as well as the relationship between them. The relationship between project management and strategic planning can best be seen from Figure 2–1. Historically, a great deal of emphasis had been placed on strategic formulation with little emphasis on strategic implementation. Now companies were recognizing that the principles of project management could be used for the implementation of strategic plans, as well as operational plans. Now, project management had the attention of senior management.

Another factor promoting project management was the acceptances of strategic business units (SBUs). There was usually less resistance to the use of project management in the SBU than in the parent company, along with greater recognition for the need to obtain horizontal as well as vertical work flow. This is shown in Figure 2–2. Project management was now recognized as a vehicle for the implementation of just about any type of plan for any type of project. Organizational charts showed project teams working horizontally across the corporation rather than vertically.

To address the far-reaching changes in the economic environment, senior managers began to ask a fundamental question: How do we plan for excellence in

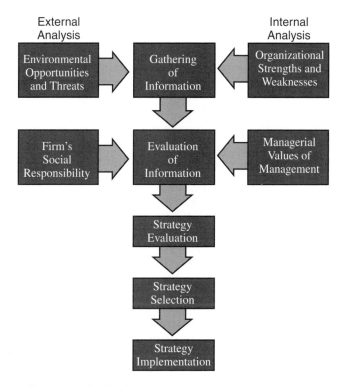

FIGURE 2–1. Basic strategic planning.

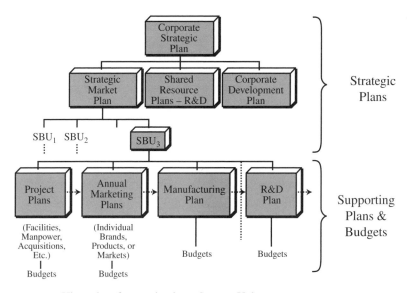

FIGURE 2–2. Hierarchy of strategic plans. *Source:* Unknown.

TABLE 2–2. STRATEGIC FACTORS IN ACHIEVING EXCELLENCE

Factor	Short-Term Applications	Long-Term Implications
Qualitative	Provide educational training Dispel illusion of a need for authority Share accountability Commit to estimates and deliverables Provide visible executive support and sponsorship	Emphasize cross-functional working relationships and team building
Organizational	De-emphasize policies and procedures Emphasize guidelines Use project charters	Create project management career path Provide project managers with reward/penalty power Use nondedicated, cross-functional teams
Quantitative	Use a single tool for planning, scheduling, and controlling	Use estimating databases

project management? In answering this question, it would be futile to expect managers to implement immediately all of the changes needed to set up modern project management in their companies. What senior managers needed was a plan expressed in terms of three broad, critical success factors: qualitative factors, organizational factors, and quantitative factors. To take advantage of the economic outlook, whatever it happened to be at a given time, senior managers needed a plan like the one shown in Table 2–2.

In the last few years, there have been new adaptations for the PMMM. These include:

- Assessing the educational maturity level of either the entire company or divisions within the company
- Identify which project management related training programs would best satisfy the companies needs
- Establishing a project management curriculum aligned with the various levels of PMMM
- Recognizing that early implementation of the project office concept can accelerate the path to maturity
- Recognizing the importance of identifying and capturing best practices in project management at each level of the PMMM
- Recognizing the importance creating a best practices library in the early levels of PMMM
- Recognizing that Six Sigma project management principles can be included in the various levels of PMMM

Current topics of discussion include the application of PMMM for use with virtual project teams and agile project teams. These topics are still in the infancy stages.

Principles of
Strategic Planning

GENERAL STRATEGIC PLANNING

Strategic planning is the process of formulating and implementing decisions about an organization's future direction. This has been shown in Figure 2–1. This process is vital to every organization's survival because it is the process by which the organization adapts to its ever-changing environment, and the process is applicable to all management levels and all types of organizations.

Let's look at the first step in strategic planning: the formulation process is the process of deciding where you want to go, what decisions must be made, and when they must be made in order to get there. It is the process of defining and understanding the business you are in and how to remain competitive within that business. The outcome of successful formulation results in the organization doing the right thing in the right way (i.e., it results in project management) by producing goods or services for which there is a demand or need in the external or internal environment. When this occurs, we say the organization has been effective as measured by market response, such as sales and market shares or internal customer acceptance. A good project management methodology can provide better customer satisfaction and a greater likelihood of repeat business. All organizations must be effective and responsive to their environments to survive in the long run.

The formulation process is performed at the top levels of the organization. Here, top management values provide the ultimate decision template for directing the course of the firm. Formulation:

- Scans the external environment and industry environment for changing conditions.
- Interprets the changing environment in terms of opportunities or threats.
- Analyzes the firm's resource base for asset strengths and weaknesses.
- Defines the mission of the business by matching environmental opportunities and threats with resource strengths and weaknesses.
- Sets goals for pursuing the mission based on top management values and sense of responsibility.

The second step in strategic planning, implementation, translates the formulated plan into policies and procedures for achieving the grand decision. Implementation involves all levels of management in moving the organization toward its mission. The process seeks to create a fit between the organization's formulated goal and its ongoing activities. Because implentation involves all levels of the organization, it results in the integration of all aspects of the firm's functioning. Integration management is a vital core competency of project management. Middle- and lower-level managers spend most of their time on implementation activities. Effective implementation results in stated objectives, action plans, timetables, policies and procedures, and results in the organization moving efficiently toward fulfillment of its mission.

WHAT IS STRATEGIC PLANNING FOR PROJECT MANAGEMENT? _____

Strategic planning for project management is the *development of a standard methodology for project management,* a methodology that can be used over and over again, and that will produce a high likelihood of achieving the project's objectives. Although strategic planning for the methodology and execution of the methodology does not guarantee profits or success, it does improve the chances of success.

One primary advantage of developing an implementation methodology is that it provides the organization with a consistency of action. As the number of interrelated functional units in organizations has increased, so have the benefits from the integrating direction afforded by the project management implementation process.

Methodologies need not be complex. Figure 3–1 shows the "skeleton" for the development of a simple project management methodology. The methodology begins with a project definition process, which is broken down into a technical baseline, a functional or management baseline, and a financial baseline. The technical baseline includes, at a minimum:

- Statement of work (SOW)
- Specifications

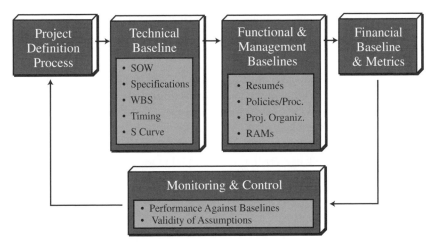

FIGURE 3–1. Methodology structuring.

- Work breakdown structure (WBS)
- Timing (i.e., schedules)
- Spending curve (S curve)

The functional or management baseline indicates how you will manage the technical baseline. This includes:

- Resumés of the key players
- Project policies and procedures
- The organization for the project
- Responsibility assignment matrices (RAMs)

The financial baseline identifies how costs will be collected and analyzed, how variances will be explained, and how reports will be prepared. Altogether, this is a simple process that can be applied to each and every project.

Without this repetitive process, subunits tend to drift off in their own direction without regard to their role as a subsystem in a larger system of goals and objectives. The objective-setting and the integration of the implementation process using the methodology assure that all of the parts of an organization are moving toward the same common objective. The methodology gives direction to diverse activities, as well as providing a common process for managing multinational projects.

Another advantage of strategic project planning is that it provides a vehicle for the communication of overall goals to all levels of management in the organization. It affords the potential of a vertical feedback loop from top to bottom, bottom to top, and functional unit to functional unit. The process of communication and its resultant understanding helps reduce resistance to change. It is extremely difficult to achieve commitment to change when employees do not understand its purpose. The strategic project planning process gives all levels an opportunity to participate, thus reducing the fear of the unknown and possibly eliminating resistance.

The final and perhaps the most important advantage is the thinking process required. Planning is a rational, logically ordered function. This is what a structured methodology provides. Many managers caught up in the day-to-day action of operations will appreciate the order afforded by a logical thinking process. Methodologies can be based upon sound, logical decisions. Figure 3–2 shows the logical decision-making process that could be part of the project selection process for an organization. Checklists can be developed for each section of Figure 3–2 to simplify the process.

The first box in Figure 3–2 is the project definition process. At this point, the project definition process simply involves a clear understanding of the objectives, which should be defined in both business and technical terms.

The second box is an analysis of the environmental situation. This includes a market feasibility analysis to determine:

- The potential size of the market for the product
- The potential risks on product liability
- The capital requirements for the product
- The market position on price
- The expected competitive response
- The regulatory climate, if applicable
- The degree of social acceptance
- Human factors (e.g., unionization)

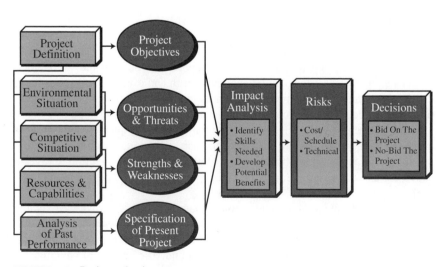

FIGURE 3–2. Project selection process.

The third box in Figure 3–2 is an analysis of the competitive situation and includes:

- The overall competitive advantage of the product
- Opportunities for technical superiority:
    - Product performance
    - Patent protection
    - Exceptional price-quality-value relationship
- Business attractiveness:
    - Type and nature of competitors
    - Structure of the competition/industry
    - Differences among competitors (price, quality, etc.)
    - Threat of substitute products
- Competitive positioning:
    - Market share
    - Rate of change in market share
    - Perceived differentiation among competitors and across various market segments
    - Positioning of the product within the product line
- Opportunities for market positioning:
    - Franchises
    - Reputation/image
    - Superior service
- Supply chain management:
    - Ownership of raw material sources
    - Vertical integration
- Physical plant opportunities:
    - Locations
    - Superior logistics support
- Financial capabilities:
    - Available capital
    - Credit rating impact
    - Wall Street support
- Efficient operations management:
    - Inventory management
    - Production
    - Distribution
    - Logistics support

The next box in Figure 3–2 is resources and capabilities. Analysis of resources and capabilities, combined with the analysis of competitive positioning just discussed, allows us to determine our strengths and weaknesses. Identifying opportunities and threats lets us identify what we *want* to do. However, it is knowing our strengths and weaknesses that lets us identify what we *can* do. Therefore, the design of any type of project management methodology must be based heavily upon what the organization can do.

Internal strengths and weaknesses can be defined for each major functional area. The design of a project management methodology can exploit the strengths in each functional area and minimize its weaknesses. Not all functional areas will possess the same strengths and weaknesses.

The following illustrates typical strengths or weaknesses for various functional organizations:

- Research and development:
    - Ability to conduct basic/applied research
    - Ability to maintain state-of-the-art knowledge
    - Technical forecasting ability
    - Well-equipped laboratories
    - Proprietary technical knowledge
    - An innovative and creative environment
    - Offensive R&D capability
    - Defensive R&D capability
    - Ability to optimize cost with performance
- Manufacturing:
    - Efficiency factors
    - Raw material availability and cost
    - Vertical integration abilities
    - Quality assurance system
    - Relationship with unions
    - Learning curve applications
    - Subsystems integration
- Finance and accounting:
    - Cash flow (present and future projections)
    - Forward pricing rates
    - Working capital requirements
- Human resource management:
    - Turnover rate of key personnel
    - Recruitment opportunities
    - Promotion opportunities
    - Having a project management career path
    - Quality of management at all levels
    - Public relations policies
    - Social consciousness
- Marketing:
    - Price-value analysis
    - Sales forecasting ability
    - Market share
    - Life cycle phases of each product
    - Brand loyalty

- Patent protection
- Turnover of key personnel

Having analyzed what we can do, we must now look at past performance to see if there are any applicable lessons learned files that could impact the current project or selection of projects. Analysis of past performance, as shown in Figure 3–2, is usually the best guide for the specifications of the present project.

The final box in Figure 3–2 is the decision on whether or not to undertake the project. This type of decision-making process is critical if we are to improve our chances of success. Historically, less than 10 percent of R&D projects ever make it through full commercialization where all costs are recovered. Part of that problem has been the lack of a structured approach for decision-making, project approval, and project execution. All this can be satisfied with a sound project management methodology.

In the absence of an explicit project management methodology, decisions are made incrementally. A response to the crisis of the moment may result in a choice that is unrelated to, and perhaps inconsistent with, the choice made in the previous moment of crisis. Discontinuous choices serve to keep the organization from moving forward. Contradictory choices are a disservice to the organization and may well be the cause of its demise. Such discontinuous and contradictory choices occur when decisions are made independently to achieve different objectives, even though everyone is supposedly working on the same project. When the implementation process is made explicit, however, objectives, missions, and policies become visible guidelines that produce logically consistent decisions.

Small companies usually have an easier time in performing strategic planning for project management excellence. Large companies with highly diversified product lines and multiple management styles find that institutionalizing changes in the way projects are managed can be very complex. Innovation and creativity in project management can be a daunting, but not impossible, task.

Effective strategic planning for project management is a never-ending effort, requiring continuous support. The two most common continuous supporting strategies are the integration opportunities strategy, outlined in Figure 3–3, and the performance improvement strategy, shown in Figure 3–4.

Figure 3–3 outlines the opportunities that exist to integrate or combine an existing methodology with other types of management approaches that may be currently in use within the company. Such other methodologies available for integration include concurrent engineering, total quality management (TQM), scope change management, and risk management. Integrated strategies provide a synergistic effect. Typical synergies include:

Project Management Process

- Tighter cost control: This results from a uniform cost reporting system in which variance reporting can be tightened and lessons learned files are maintained and updated.

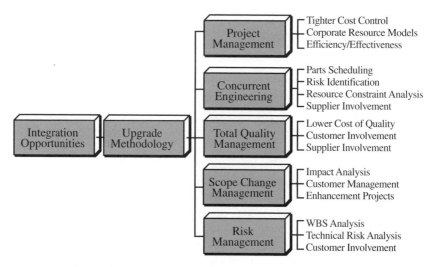

FIGURE 3–3. Integration opportunities between process strategies.

- Corporate resource models: Companies are now able to develop total company resource models and capacity planning models to determine how efficiently the existing resources are being utilized and how much new business can be undertaken.
- Efficiency/effectiveness: A good methodology allows for the capturing and comparison of metrics to show that the organization is performing

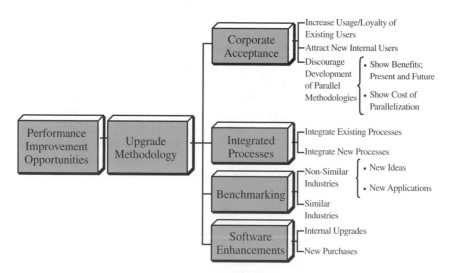

FIGURE 3–4. Qualitative process improvement opportunities.

more work in less time and with fewer resources. Such data verifies the existence of economies of scale.

Concurrent Engineering Process

- Parts scheduling: Improvement can be made in the way that parts are ordered and tracked. As an organization overlaps activities to compress the schedule, timely delivery of materials is crucial.
- Risk identification: Overlapping activities increase the risks on a project. Better risk management practices are essential.
- Resource constraint analysis: Overlapping activities during concurrent engineering require that sufficient resources be available. Models are available to define the resource constraints and recommend ways to deal with limited availability of resources.
- Supplier involvement: Overlapping activities not only increase your risks but can also increase the risks for your supplier. A good methodology allows for better customer interfacing.

Total Quality Management (TQM)

- Lower cost of quality: Many of the basic principles of project management are also the basic principles of TQM. A good project management methodology will allow for the maximization of benefits for both.
- Customer involvement: A good methodology allows you to get closer to your customers. This could easily result in customer involvement in ways to improve quality for both products and services.
- Supplier involvement: A good methodology allows you to get closer to your supplier base. Suppliers will often come up with ideas to improve quality, thus solidifying your relationship with them. They also may have information from other companies they supply, possibly even your competitors, and may be willing to release this information.

Scope Change Management

- Impact analysis: A good methodology allows for checklists and forms for accurately determining the time, cost, and quality impact resulting from scope changes. It also puts in place a regimented process for scope change management.
- Customer management: Customers want to believe that all changes are no-cost changes (to the customer), and that the changes can be made at any time and in any life cycle phase. A good methodology that completely outlines the change management process allows for better customer management.
- Enhancement projects: A good project management methodology allows for a clear distinction as to whether the change should be made now or possibly later as an enhancement project. It addresses the question of how imperative the change actually is.

Risk Management

- WBS analysis: A good methodology provides guidelines on how deep into the WBS risk analysis should be performed.

- Technical risk analysis: Risk analysis is reasonably well defined for schedule and cost risks. Very little is known about technical risk management. A good methodology may provide templates on how to perform technical risk management.
- Customer involvement: Your firm's perception of risks, specifically which risks are worth taking and which are not, may be significantly different than your customer's perceptions. Customer involvement in risk analysis is essential.

The qualitative process improvement strategy shown in Figure 3–4 is designed to improve the efficiency of the existing methodology and to find new applications for it. The integrated process strategies of Figure 3–3 are one part of these process improvement strategies, as shown. Process improvement is discussed further in Chapter 9.

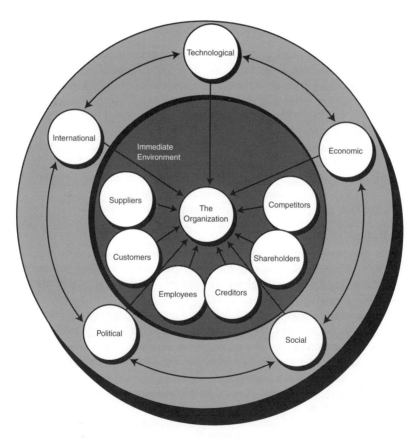

FIGURE 3–5. The macroenvironment of business. *Note:* Stakeholders are identified in the "immediate environment" circle. *Source:* From *The Changing Environment of Business, A Managerial Approach,* 4th Edition, by G. Starling © 1995. Reprinted with permission of South-Western, a division of Thomson Learning: www.thomsonrights.com. Fax 800 730-2215.

The goal of most organizations is to be more profitable than their competitors. Project management methodologies contribute to profitability through more efficient execution of the project and implementation of the methodology. This is another valid reason mandating continuous strategic planning.

A good project management methodology will be responsive to all environmental factors and will serve all of its stakeholders. Stakeholders are people who have a vested interest in the company's performance and who have claims on its performance. Figure 3–5 shows, as part of the immediate environment, six commonly used categories of stakeholder: suppliers, customers, employees, creditors, shareholders, and even competitors. Some organizations would also identify government officials and society at large as being among their multiple stakeholders. One part of strategic planning for project management may include prioritizing the order in which stakeholders will be satisfied if and when a problem exists. A good project management methodology may also include a "standard practices" section, which will discuss moral and ethical considerations involved in dealing with stakeholders.

EXECUTIVE INVOLVEMENT

Senior management's involvement in strategic planning is essential if the process is to move ahead quickly and if full employee commitment and acceptance is to be achieved. The need for involvement is essential:

- A visible general endorsement is mandatory.
- An executive champion (not sponsor) must be assigned.
- The executive champion must initiate the process.
- The executive champion must make sure that the ideas/aspirations of senior management are included throughout the methodology.
- The executive champion must verify the validity of the corporate assumptions, including:
  - Forward pricing rate data
  - Targeted customers/industries
  - Reporting requirement for senior management
  - Strategic trends
  - Customer interfacing requirements

If senior management's support is not visible right from the onset, then:

- The workers may believe that senior management is not committed to the process.
- Functional managers may hesitate to provide valuable support, believing that the process is unreal.
- The entire process may lack realism and waste time.

Another critical function of senior management is determining "strategic timing." A strategic plan is a timed sequence of conditional moves that involve the deployment of resources. The executive champion must either develop or approve the strategic timing activities, which include:

- Establishing the timetable for major moves
- Establishing resource requirements and assuring availability
- Providing funding release time for critical assets and hardware/software purchases to support the project management systems

THE GENERAL ENVIRONMENT

The first step in any strategic planning process is an understanding of the general environment in which the strategy must be targeted and executed (refer again to Figure 3–5). Historically, the general environment includes:

- The demographic segment
- The economic segment
- The political/legal segment
- The sociocultural segment
- The technological segment

In general strategic planning, these segments are heavily oriented toward the external environment. For project management, the focus is more internal than external.

The Demographic Segment
For general strategic planning, we focus on such factors as population size, age structure, geographic structure, ethnic mix, and income distribution. For project management, the focus is more internal. The factors we look at will include:

- Corporate size: How many functional units will use the methodology? Will there be pockets of use or corporate-wide acceptance?
- Age structure: What will be the average age of the users of the methodology? Age structure can affect both risk-taking and willingness/ability to work overtime.
- Geographic dispersion: If the firm is multinational, how do we get everyone to support the methodology? Will there exist language/communication complexities?
- Types of projects: Will the methodology be general enough for all types of projects, or will we need multiple methodologies?

The Economic Segment

For general strategic planning, the economic environment is the external economic environment and how it affects the operations of the firm. Included in this segment would be inflation rates, interest rates, trade budget/surpluses, personal saving rates, business saving rates and gross domestic product. For project management strategies, the economic environment will include:

- Cost of capital: How much will it cost us to borrow money for new product development or on an interim basis to account for cash flow deficits?
- Forward pricing rates: Based upon current knowledge, what will our costs look like over the next several years?
- Quality of cost estimates: How accurate are our cost estimates, and are there lessons learned?

The Political/Legal Segment

For general strategic planning, the political/legal segment includes laws on antitrust, taxation, and labor training, and philosophies on deregulation and education. For project management strategic planning, the list could include the above items for multinational efforts, but generally includes:

- Customer-interfacing: This includes the development of a standard practice manual on morality and ethics in dealing with customers. It could also include a corporate "credo" that specifies that the best interests of the customer come first.
- Product liability/truth of disclosure: Do we have supporting data for information presented to the customers or consumers?
- Changing laws: Does our methodology allow for changes if new laws are enacted?

The Sociocultural Segment

The sociocultural segment generally includes topics such as women/minorities in the workforce, quality of work life, environmental concerns, and career preferences. For strategic planning for project management, the list would include:

- Customer requirements: Do our customers mandate the hiring of women/minorities on our projects? Do the customers require that our subcontracts go to union shop organizations only?
- Health/safety issues: Does our methodology specify that we do not expect the employees to violate health and safety regulations?
- Overtime: How much overtime are employees expected to perform? This includes both exempt and nonexempt employees.
- Career path: Is project management regarded as a career path position?

The Technological Segment

The technological segment is basically the same for general strategic planning and strategic planning for project management. Included in this segment we have:

- Offensive technology: Do we have the skill to develop new products and, if so, does the methodology account for technical risk-taking in this regard?
- Defensive technology: How quickly and effectively can we defend our existing products through enhancements? Does our methodology allow for shortcuts for enhancement projects?
- Purchasing of technology: Does our company allow us to purchase technology (hardware, software, etc.) to improve our management processes?
- Technology gap: Does a technology gap exist between us and our competitors? Does the methodology allow for risk-taking to close the gap?
- The freedom to innovate: Is the methodology rigidly structured or does it allow some degree of freedom for creativity?

CRITICAL SUCCESS FACTORS FOR STRATEGIC PLANNING

Critical success factors for strategic planning for project management include those activities that must be performed if the organization is to achieve its long-term objectives. Most businesses have only a handful of critical success factors. However, if even one of them is not executed successfully, the business's competitive position may be threatened.

The critical success factors in achieving project management excellence apply equally to all types of organizations, even those that have not fully implemented their project management systems. Though most organizations are sincere in their efforts to fully implement their systems, stumbling blocks are inevitable and must be overcome. Here's a list of common complaints from project teams:

- There's scope creep in every project and no way to avoid it.
- Completion dates are set before project scope and requirements have been agreed upon.
- Detailed project plans identifying all of the project's activities, tasks, and subtasks are not available.
- Projects emphasize deadlines. We should emphasize milestones and quality and not time.
- Senior managers don't always allow us to use pure project management techniques. Too many of them are still date-driven instead of requirements-driven. Original target dates should be used only for broad planning.
- Project management techniques from the 1960s are still being used on most projects. We need to learn how to manage from a plan and how to use shared resources.

- Sometimes we are pressured to cut estimates low to win a contract, but then we have to worry about how we'll accomplish the project's objectives.
- There are times when line personnel not involved in a project change the project budget to maintain their own chargeability. Management does the same.
- Hidden agendas come into play. Instead of concentrating on the project, some people are out to set precedents or score political points.
- We can't run a laboratory without equipment, and equipment maintenance is a problem because there's no funding to pay for the materials and labor.
- Budgets and schedules are not coordinated. Sometimes we have spent money according to the schedule but are left with only a small percentage of the project activities complete.
- Juggling schedules on multiple projects is sometimes almost impossible.
- Sometimes we filter information from reports to management because we fear sending them negative messages.
- There's a lot of caving in on budgets and schedules. Trying to be a good guy all the time is a trap.

With these comments in mind, let's look at the three critical success factors in achieving project management excellence: qualitative, organizational, and quantitative factors.

QUALITATIVE FACTORS

If excellence in project management means a continuous stream of successfully completed projects, then our first step should be to define *success*. As discussed in Chapter 1, success in projects has traditionally been defined as achieving the project's objectives within the following constraints:

- Allocated time
- Budget cost
- Desired performance at technical or specification level
- Quality standards as defined by customers or users

In experienced organizations, the four preceding parameters have been extended to include the following:

- With minimal or mutually agreed upon scope changes
- Without disturbing the organization's corporate culture or values
- Without disturbing the organization's usual work flow

These last three parameters deserve further comment.

Organizations that eventually achieve excellence are committed to quality and up-front planning so that minimal scope changes are required as the project progresses. Those scope changes that are needed must be approved jointly by both the customer and the contractor. A well-thought-out process for handling scope changes is in place in such organizations. Even in large profit-making, project-driven industries, such as aerospace, defense, and large construction, tremendous customer pressure can be expected to curtail any "profitable" scope changes introduced by the contractor.

Most organizations have well-established corporate cultures that have taken years to build. On the other hand, project managers may need to develop their own subcultures for their projects, particularly when the projects will require years to finish. Such temporary project cultures must be developed within the limitations of the larger corporate culture. The project manager should not expect senior officers of the company to allow the project manager free rein.

The same limitations affect organizational work flow. Most project managers working in organizations that are only partially project-driven realize that line managers in their organizations are committed to providing continuous support to the company's regular functional work. Satisfying the needs of time-limited projects may only be secondary. Project managers are expected to keep the welfare of their whole companies in mind when they make project decisions.

> For companies to reach excellence in project management, executives must learn to define project success in terms of both what is good for the project and what is good for the organization.

Executives can support project managers by reminding them of this two-part responsibility by:

- Encouraging project managers to take on nonproject responsibilities, such as administrative activities
- Providing project managers with information on the company's operations and not just information pertaining to their assigned projects
- Supporting meaningful dialogue among project managers
- Asking whether decisions made by project managers are in the best interest of the company as a whole

ORGANIZATIONAL FACTORS

Coordination of organizational behavior in project management is a delicate balancing act, something like sitting on a bar stool. Bar stools usually come with three legs to keep them standing. So does project management: one is the project manager, one is the line manager, and one is the project sponsor. If one of the legs is lost or unusable, the stool will be very difficult to balance.

Although line managers are the key to successful project management, they will have a lot of trouble performing their functions without effective interplay with the project's manager and corporate sponsor. In unsuccessful projects, the project manager has often been vested with power (authority) over the line managers involved. In successful projects, project and line managers are more likely to have shared authority. The project manager will have negotiated the line managers' commitment to the project and worked through them, not around them. The project manager probably provided recommendations regarding employee performance. And leadership was centered around the whole project team, not just the project manager.

In successful project management systems, the following equation always holds true:

$$Accountability = Responsibility + Authority$$

When project and line managers view each other as equals, they share equally in the management of the project, and thus they share equally the authority, responsibility, and accountability for the project's success. Obviously the sharing of authority makes shared decision-making easier. The project management methodology must account for shared accountability. A few suggestions for executive project sponsors follow:

- Do not increase the authority of the project manager at the expense of the line managers.
- Allow line managers to provide technical direction to their people, if at all possible.
- Encourage line managers to provide realistic time and resource estimates, and then work with the line managers to make sure they keep their promises.
- Above all, keep the line managers fully informed.

In organizations that have created effective project management systems, the role of the executive manager has changed along with project management. Early in the implementation of project management, executives were actively involved in the everyday project management process. But as project management has come into its own and as general economic conditions have changed, executive involvement has become more passive, and project sponsors now usually concentrate on long-term and strategic planning. They have learned to trust project managers to make the day-to-day decisions and they have come to view project management as a central factor in their company's success.

Project sponsors provide visible, ongoing support. Their role is to act as a bodyguard for the project and the project manager. Unlike other executives on the senior management team, individual project sponsors may play a more active role in projects, depending on how far along the project is. Early in the project's functioning, for example, the project sponsor might help the project manager define the project's requirements. Once that is done, the sponsor resumes a less active role and receives project information only as needed.

In successful project management systems that carry a high volume of ongoing project work, an executive sponsor may not be assigned to low-dollar-value or low-priority projects. Middle managers may fill the sponsorship role in some cases. But no matter what the size or value of the project, project sponsors today are responsible for the welfare of all members of their project teams, not just that of the project manager.

The existence of a project sponsor implies visible, ongoing executive support for project management. And executive support motivates project personnel to excel. Executive project sponsorship also supports the development of an organizational culture that fosters confidence in the organization's project management systems.

Conclusion: Executive project sponsorship must exist and be visible so that the project–line manager interface is in balance.

Recommendations for obtaining maturity include:

- Educate the executives as to the benefits of project management.
- Convince the executives of the necessity for ongoing, visible support in the capacity of a project sponsor.
- Convince executives that they need not know all the details. Provide them with the least information that tells the most.

QUANTITATIVE FACTORS

The third factor in achieving excellence in project management is the implementation and acceptance of project management tools to support the methodology. (See the discussion of project management tools in Chapter 4.) Some companies are quick to implement network-based scheduling tools such as PERT, CPM and Precedence Networks, but many are reluctant to accept other mainframe or personal computer network software for project planning, project cost estimating, project cost control, resource scheduling, project tracking, project audits, or project management information systems.

Mainframe project management tools have been resurrected in the past few years. These new mainframe products are being used mainly for total company project control. However, some executives have been slow to accept these sophisticated tools. The reasons for this are:

- Upper management may not like the reality of the output.
- Upper management may prefer to use their own techniques rather than the system for the planning, budgeting, and decision-making process.
- Day-to-day planners may not use the packages for their own projects.
- Upper management may not demonstrate support and commitment to training.
- Use of sophisticated mainframe packages requires strong internal communication lines for support .
- Clear, concise reports may be lacking even though report generators exist.
- Mainframe packages do not always provide for immediate turnaround of information.
- The business entity may not have any project management standards in place prior to implementation.
- Implementation may highlight middle management's inexperience in project planning and organizational skills.
- Sufficient/extensive resources required (staff, equipment, etc.) may not be in place.
- Business environment and organizational structure may not be appropriate to meet project management/planning needs.
- Software utilization training without project management training is insufficient.
- Software may be used inappropriately as a substitute for the extensive interpersonal and negotiation skills required by project management.
- The business entity may not have predetermined the extent and appropriate use of the software within the organization.

Conclusions: Project management education must precede software education. Also, executives must provide the same encouragement and support for the use of the software as they do for project management.

The following recommendations can help accelerate the maturity process:

- Educating people in the use of sophisticated software and having them accept its use is easier if the organization is already committed to project management.
- Executives must provide standards and consistency for the information they wish to see in the output.
- Executive knowledge (overview) in project management principles is necessary to provide meaningful support.
- Not everyone needs to become an expert in the use of the system. One or two individuals can act as support resources for multiple projects.

IDENTIFYING STRATEGIC RESOURCES _____

All businesses have corporate competencies and resources that distinguish them from their competitors. These competencies and resources are usually identified in terms of a company's strengths and weaknesses. Deciding upon what a company *should* do can only be achieved after assessing the strengths and weaknesses to determine what the company *can* do. Strengths support windows of opportunities, whereas weaknesses create limitations. What a company can do is based upon the quality of its resources.

Strengths and weaknesses can be identified at all levels of management. Senior management may have a clearer picture of the overall company's position in relation to the external environment, whereas middle management may have a better grasp of the internal strengths and weaknesses. Unfortunately, most managers do not think in terms of strengths and weaknesses and, as a result, they worry more about what they *should* do than about what they *can* do.

Although all organizations have strengths and weaknesses, no organization is equally strong in all areas. Procter & Gamble, Budweiser, Coke, and Pepsi are all known for their advertising and marketing. Computer firms are known for technical strengths, whereas General Electric has long been regarded as the training ground for manufacturing executives. Large firms have vast resources with strong technical competency, but they often react slowly when change is needed. Small firms can react quickly but have limited strengths. Any organization's strengths and weaknesses can change over time and must, therefore, be closely monitored.

Strengths and weaknesses are internal measurements of what a company can do and assessment of them must be based upon the quality of the company's resources. Consider the situation shown in Figure 3–6. Even a company with a world-class methodology in project management will not be able to close the performance gap in Figure 3–6 until the proper internal or subcontracted resources are available. Methodologies, no matter how good, are executed by use of resources. Project management methodologies do not guarantee success. They simply increase the chances for success provided that (1) the project objective is realistic and (2) the proper resources are available along with the skills needed to achieve the objective.

Tangible Resources

In basic project management courses, the strengths and weaknesses of a firm are usually described in the terms of its tangible resources. The most common classifications for tangible resources are:

- Equipment
- Facilities
- Manpower
- Materials
- Money
- Information/technology

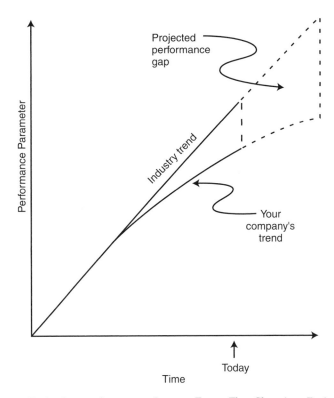

FIGURE 3–6. Projecting performance. *Source:* From *The Changing Environment of Business, A Managerial Approach,* 4th Edition, by G. Starling © 1995. Reprinted with permission of South-Western, a division of Thomson Learning: www.thomsonrights.com. Fax 800 730-2215.

Another representation of resources is shown in Figure 3–7. Unfortunately, these crude types of classification do not readily lend themselves to an accurate determination of internal strengths and weaknesses for project management. A more useful classification would be human resources, nonhuman resources, organizational resources, and financial resources.

Human Resources

Human resources are the knowledge, skills, capabilities, and talent of the firm's employees. This includes the board of directors, managers at all levels, and employees as a whole. The board of directors provides the company with considerable experience, political astuteness, and connections, and possibly sources of borrowing power. The board of directors is primarily responsible for selecting the CEO and representing the best interest of the diverse stakeholders as a whole.

Top management is responsible for developing the strategic mission and making sure that the strategic mission satisfies the shareholders. All too often, CEOs have singular strengths in only one area of business, such as marketing, finance, technology, or production.

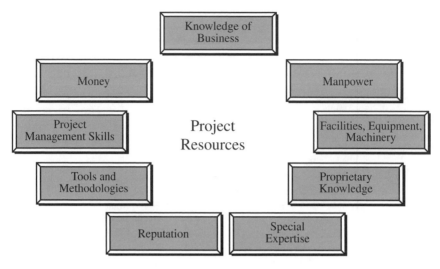

FIGURE 3–7. Project resources.

The biggest asset of senior management is its decision-making ability, especially during project planning. Unfortunately, all too often senior management will delegate planning (and the accompanying decision-making process) to staff personnel. This may result in no effective project planning process within the organization and may lead to continuous replanning efforts.

Another important role of senior management is to define clearly its own managerial values and the firm's social responsibility (see Figure 2–1). A change in senior management could result in an overnight change in the organization's managerial values and its definition of its social responsibility. This could require an immediate update of the firm's project management methodology.

Lower and middle management are responsible for developing and maintaining the "core" technical competencies of the firm. Every organization maintains a distinct collection of human resources. Middle management must develop some type of cohesive organization such that synergistic effects will follow. It is the synergistic effect that produces the core competencies that lead to sustained competitive advantages and a high probability of successful project execution.

Nonhuman Resources

Nonhuman resources are physical resources that distinguish one organization from another. Boeing and IBM both have sustained competitive advantages but have different physical resources. Physical resources include plant and equipment, distribution networks, proximity of supplies, availability of a raw material, land, and labor.

Companies with superior nonhuman resources may not have a sustained competitive advantage without also having superior human resources. Likewise, a company with strong human resources may not be able to take advantage of windows of opportunity unless it also has strong physical resources. An Ohio-based company had a 30-year history of sustained competitive advantage on R&D projects that were won through competitive bidding. As times changed however, senior management saw that the potential for megaprofits now lay in production. Unfortunately, in order to acquire the resources needed for physical production, the organization diluted some of its technical resources. The firm learned a hard lesson in that the management of human resources is not the same as the management of nonhuman resources. The firm also had to reformulate its project management methodology to account for manufacturing operations.

Firms that endeavor to develop superior manufacturing are faced with two critical issues. First, how reliable are the suppliers? Do the suppliers maintain quality standards? Are the suppliers cost effective? The second concern, and perhaps the more serious of the two, is the ability to cut costs quickly and efficiently to remain competitive. This usually leads to some form of vertical integration.

Organizational Resources
Organizational resources are the glue that holds all of the other resources together. Organizational resources include the organizational structure, the project office, the formal (and sometimes informal) reporting structure, the planning system, the scheduling system, the control system, and the supporting policies and procedures. Decentralization can create havoc in large firms where each strategic business unit (SBU), functional unit, and operating division can have its own policies, procedures, rules, and guidelines. Multiple project management methodologies can cause serious problems if resources are shared between SBUs.

Financial Resources
Financial resources are the firm's borrowing capability, credit lines, credit rating, ability to generate cash, and relationship with investment bankers. Companies with quality credit ratings can borrow money at a lower rate than companies with nonquality ratings. Companies must maintain a proper balance between equity and credit markets when raising funds. A firm with strong, continuous cash flow may be able to fund growth projects out of cash flow rather than through borrowing. This is the usual financial-growth strategy for a small firm.

Intangible Resources
Human, physical, organizational, and financial resources are regarded as tangible resources. There are also intangible resources that include the organizational culture, reputation, brand name, patents, trademarks, know-how, and relationships with customers and suppliers. Intangible resources do not have the visibility that tangible resources possess, but they can lead to a sustained competitive advantage. When companies develop a "brand name," it is nurtured through advertising and marketing and is often accompanied by a slogan. Project management methodologies can include paragraphs on how to protect the corporate image or brand name.

Social Responsibility

Social responsibility is also an intangible asset, although some consider it both intangible and tangible. Social responsibility is the expectation that the public perceives that a firm will make decisions that are in the best interest of the public as a whole. Social responsibility can include a broad range of topics from environmental protection to consumer safeguards to consumer honesty and employing the disadvantaged. An image of social responsibility can convert a potential disaster into an advantage. Johnson & Johnson, for example, earned high marks for social responsibility in the way it handled the two Tylenol tragedies in the 1980s. Nestlé, on the other hand, earned low marks for its role in the infant-formula controversy.

WHY DOES STRATEGIC PLANNING FOR PROJECT MANAGEMENT SOMETIMES FAIL?

We have developed a strong case earlier for the benefits of strategic planning for project management. Knowledge about this process is growing, and new information is being disseminated rapidly. Why, then, does this process often fail? Following are some of the problems that can occur during the strategic planning process. Each of these pitfalls must be considered carefully if the process is to be effective.

- Lack of CEO endorsement: Any type of strategic planning process must originate with senior management. They must start the process and signal their own aspirations. A failure by senior management to endorse strategic planning may signal line management that the process is unreal.
- Failure to reexamine: Strategic planning for project management is not a one-shot process. It is a dynamic, continuous process of reexamination, feedback, and updating.
- Being blinded by success: Simply because a few projects are completed successfully does not mean that the methodology is correct, nor does it imply that improvements are not possible. A belief that "you can do no wrong" usually leads to failure.
- Overresponsiveness to information: Too many changes in too short a time frame may leave employees with the impression that the methodology is flawed or that its use may not be worth the effort. The issue to be decided here is whether changes should be made continuously or at structured time frames.
- Failure to educate: People cannot implement successfully and repetitively a methodology they do not understand. Training and education on the use of the methodology is essential.
- Failure of organizational acceptance: Company-wide acceptance of the methodology is essential. This may take time to achieve in large organi-

Marketing

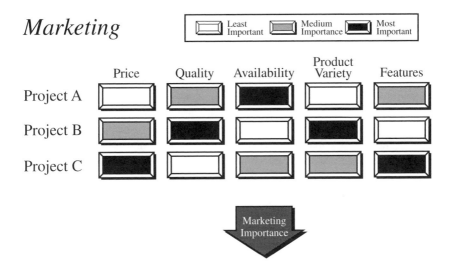

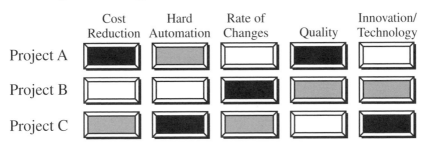

Manufacturing

FIGURE 3–8. Differences in strategic importance.

zations. Strong, visible executive support may be essential for rapid acceptance.

- Failure to keep the methodology simple: Simple methodologies based upon guidelines are ideal. Unfortunately, as more and more improvements are made, there is a tendency to go from informality using guidelines to formality using policies and procedures.

- Blaming failures on the methodology: Project failures are not always the result of poor methodology; the problem may be poor implementation. Unrealistic objectives or poorly defined executive expectations are two common causes of poor implementation. Good methodologies do not guarantee success, but they do imply that the project will be managed correctly.

- Failure to prioritize: Serious differences can exist in the importance that different functional areas, such as marketing and manufacturing, assign to strategic project objectives. Figure 3–8 shows three projects and how they are viewed differently by marketing and manufacturing. A common, across-company prioritization system may be necessary.
- Rapid acquisitions: Sometimes an organization will purchase another company as part of its long-term strategy for vertical integration. Backward integration occurs when a firm purchases suppliers of components or raw materials in order to reduce its dependency on outside sources. Forward integration occurs when an organization purchases the forward channels of distribution for its products. In either case, the company's projects will now require more work, and this must be accounted for in the methodology. Changes may occur quickly.

Only by watching out for these potential problems can a firm hope to avoid them (or at least to minimize their negative effects). This is the path to success in strategic planning for project management.

An Introduction to the Project Management Maturity Model (PMMM)

INTRODUCTION

All companies desire to achieve maturity and excellence in project management. Unfortunately, not all companies recognize that the timeframe can be shortened by performing strategic planning for project management. The simple use of project management, even for an extended period of time, does *not* necessarily lead to excellence. Instead, it can result in repetitive mistakes and, what's worse, learning from your own mistakes rather than from the mistakes of others.

Strategic planning for project management is unlike other forms of strategic planning in that it is most often performed at the middle-management level, rather than by executive management. Executive-level management is still involved, mostly in a supporting role, and provides funding together with employee release time for the effort. Executive involvement will be necessary to make sure that whatever is recommended by middle management will not result in unwanted changes to the corporate culture.

Although executives may not be active participants in or champions of the project management process, they must still drive the development and implementation process from the top down. In one automotive supplier, middle management developed an outstanding project management methodology. Senior management sponsored the implementation process to make sure that the entire organization bought into the methodology and used it. After implementation, executive sponsorship diminished. This resulted in a very weak continuous improvement process because nobody was driving the change process from the top down. An executive champion was then reinstated, and continuous improvement flourished to the point where this supplier now has one of the best project management methodologies within the automotive industry.

Organizations tend to perform strategic planning for new products and services by laying out a well-thought-out plan and then executing the plan with the precision of a surgeon. Unfortunately, strategic planning for project management, if performed at all, is done on a trial-by-fire basis. However, there are models that can be used to assist corporations in performing strategic planning for project management and achieving maturity and excellence in a reasonable period of time.

THE FOUNDATION FOR EXCELLENCE

The foundation for achieving excellence in project management can best be described as the project management maturity model (PMMM), which is comprised of five levels, as shown in Figure 4–1. Each of the five levels represents a different degree of maturity in project management. Each level is discussed in detail in the remaining chapters. The levels are:

- Level 1—Common language: In this level, the organization recognizes the importance of project management and the need for a good understanding of the basic knowledge on project management and the accompanying language/terminology. Not all companies agree on project management terminology. The terminology used in the Project Management Body of Knowledge (PMBOK®) Guide is not the only acceptable termi-

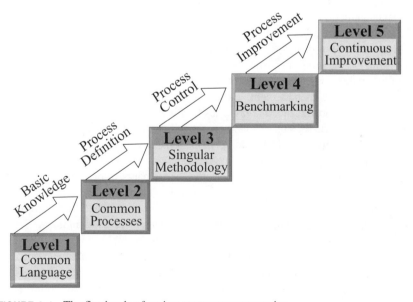

FIGURE 4–1. The five levels of project management maturity.

nology. Many companies that are quite successful in project management have their own terminology.

- Level 2—Common processes: In this level, the organization recognizes that common processes need to be defined and developed such that successes on one project can be repeated on other projects. Also included in this level is the recognition of the application and support of the project management principles to other methodologies employed by the company.
- Level 3—Singular methodology: In this level, the organization recognizes the synergistic effect of combining all corporate methodologies into a singular methodology, the center of which is project management. The synergistic effects also make process control easier with a single methodology than with multiple methodologies.
- Level 4—Benchmarking: This level contains the recognition that process improvement is necessary to maintain a competitive advantage. Benchmarking must be performed on a continuous basis. The company must decide whom to benchmark and what to benchmark.
- Level 5—Continuous improvement: In this level, the organization evaluates the information obtained through benchmarking and must then decide whether or not this information will enhance the singular methodology.

When we talk about levels of maturity (and even life cycle phases), there exists a common misbelief that all work must be accomplished sequentially (i.e., in series). This is not necessarily true. Certain levels can and do overlap. The magnitude of the overlap is based upon the amount of risk the organization is willing to tolerate. For example, a company can begin the development of project management checklists to support the methodology while it is still providing project management training for the workforce. A company can create a center of excellence (COE) in project management or a project management office (PMO) before benchmarking is undertaken.

OVERLAP OF LEVELS

Although overlapping does occur, the order in which the phases are completed cannot change. For example, even though Level 1 and Level 2 can overlap, Level 1 *must* still be completed before Level 2 can be completed. Overlapping of several of the levels can take place, as shown in Figure 4–2.

- Overlap of Level 1 and Level 2: This overlap will occur because the organization can begin the development of project management processes either while refinements are being made to the common language or during training.
- Overlap of Level 3 and Level 4: This overlap occurs because, while the organization is developing a singular methodology, plans are being made as to the process for improving the methodology.

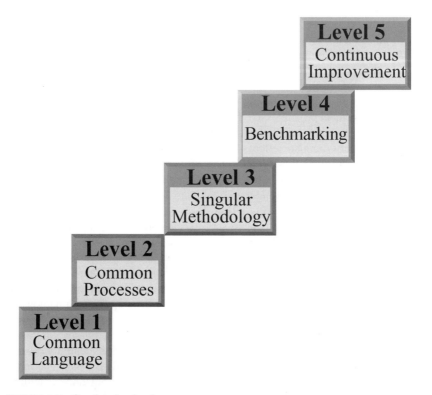

FIGURE 4–2. Overlapping levels.

- Overlap of Level 4 and Level 5: As the organization becomes more and more committed to benchmarking and continuous improvement, the speed by which the organization wants changes to be made can cause these two levels to have significant overlap. The feedback from Level 5 back to Level 4 and Level 3, as shown in Figure 4–3, implies that these three levels form a continuous improvement cycle, and it may even be possible for all three of these levels to overlap.

Level 2 and Level 3 generally do not overlap. It may be possible to begin some of the Level 3 work before Level 2 is completed, but this is highly unlikely. Once a company is committed to a singular methodology, work on other methodologies generally terminates.

Also, if a company is truly astute in project management, it may be possible to begin benchmarking efforts even as early as Level 1. This way the company may learn from the mistakes of others rather than from its own mistakes. It is possible for Level 4 to overlap all of the first three levels.

For example, a company recognized that project management would certainly be beneficial and began performing strategic planning for project manage-

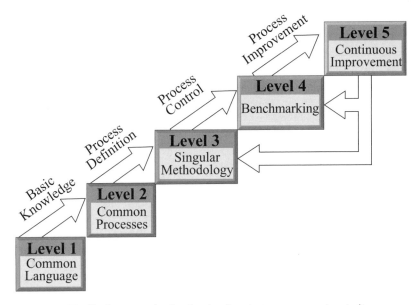

FIGURE 4–3. Feedback among the five levels of project management maturity.

ment. Senior management approved the creation of a project management office (i.e., a component of Level 4 of the PMMM) to head up the strategic planning for project management effort. The project management office (PMO) began functioning while the company began activities in Level 1 of the PMMM. In less than six months, the company was able to complete the first three levels of the PMMM.

RISKS

Risks can be assigned to each level of the PMMM. For simplicity's sake, the risks can be labeled as low, medium, and high. The level of risk is most frequently associated with the impact of having to change the corporate culture. Assigning risk is a subjective assessment on the way that the corporate culture might react at a specific level of the PMMM. The following definitions can be assigned to these three risks:

- Low risk: There will be virtually no impact on the corporate culture, or the corporate culture is dynamic and readily accepts change.
- Medium risk: The organization recognizes that change is necessary but may be unaware of the impact of the change. Instituting multiple-boss reporting would be an example of a change carrying medium risk.

- High risk: High risks occur when the organization recognizes that the changes resulting from the implementation of project management will cause a change in the corporate culture. Examples include the creation of project management methodologies, policies, and procedures, as well as decentralization of authority and decision-making.

Level 3 has the highest risks and degree of difficulty for the organization. This is shown in Figure 4–4. Once an organization is committed to Level 3, the time and effort needed to achieve the higher levels of maturity have a low degree of difficulty. Achieving Level 3, however, may require a major shift in the corporate culture.

Everyone in the organization, perhaps even on a global basis, must use the same methodology for project management. This could easily lead to changes in work habits, social groups, and comfort zones. Unfortunately, developing a singular methodology, often referred to as an enterprise project management methodology, is not easy. Most people argue that two methodologies are needed: one for information systems projects and another for new product development. Some companies that claim that they are using a singular enterprise project management methodology may have the information systems methodology as a subset of the new product development methodology.

Implementing an enterprise methodology multinationally creates additional problems. Each country can have its own laws, types of contracts, labor and employment requirements, and standards. Language barriers are also an issue.

The following chapters have detailed descriptions of each of the five levels of the PMMM. For each of the five levels of maturity, we discuss:

- The characteristics of the level
- What roadblocks exist that prevent us from reaching the next level

Level	Description	Degree of Difficulty
1	Common Language	Medium
2	Common Processes	Medium
3	Singular Methodology	High
4	Benchmarking	Low
5	Continuous Improvement	Low

FIGURE 4–4. Degree of difficulty associated with each level of the PMMM.

- What must be done to reach the next level
- Potential risks

ASSESSMENT INSTRUMENTS _____

Also included in each of the next five chapters is an assessment instrument to help you determine your organization's degree of maturity at each level. No two companies implement project management the same way. Since maturity will be different from company to company, the questions in these assessments can be modified to satisfy the needs of individual companies. Simply stated, using the principles contained in each chapter, you can customize the assessment instruments for each level. Not all companies follow the PMBOK® Guide. And those that do follow it, do not necessarily emphasize all areas of the PMBOK® Guide.

The assessment instruments can be custom-designed to include company-specific areas of maturity such as supply chain management, the project management office, and portfolio management implementation. The assessment grading system can also be custom-designed to identify ways to speed up the implementation process. For additional information on customization, contact Lori Milhaven at the International Institute for Learning, 212-515-2121.

Level 1: Common Language

INTRODUCTION

Level 1 is the level in which the organization first recognizes the importance of project management. The organization may have a cursory knowledge of project management or simply no knowledge at all. There are certain characteristics of Level 1, as shown in Figure 5–1:

- If the organization is using project management at all, the use is sporadic. Both senior management and middle-level management provide meaningless or "lip service" support to the use of project management. Executive-level support is nonexistent.
- There may exist small "pockets" of interest in project management, with most of the interest existing in the project-driven areas of the firm.
- No attempt is made to recognize the benefits of project management. Managers are worried more about their own empires, power, and authority, and appear threatened by any new approach to management.
- Decision-making is based upon what is in the best interest of the decision-maker, rather than the firm as a whole.
- There exists no investment or support for project management training and education for fear that this new knowledge may alter the status quo.

In Level 1, project management is recognized, as in all companies, but *not* fully supported. There is resistance to change, and some companies never get beyond this level.

Common Language

- Lip Service to Project Management
- Virtually No Executive-Level Support
- Small "Pockets" of Interest
- No Attempt to Recognize the Benefits of Project Management
- Self-Interest Comes Before Company's Best Interest
- No Investment in Project Management Training and Education

FIGURE 5–1. Characteristics of Level 1.

The starting point to overcome the characteristics of Level 1 is a sound, basic knowledge of the principles of project management. Education is the "name of the game" to complete Level 1. Educational programs on project management cover the principles of project management, advantages (and disadvantages) of project management methodologies, and the basic language of project management.

Project management certification training courses are ideal to fulfill the organizational needs to reach Level 1 of the project management maturity model (PMMM). Project management and total quality management (TQM) are alike in that both require an all-employee training program that begins at the senior levels of management. However, the magnitude of the training program and the material covered can vary, based upon the type of employees, skills needed, and the size and nature of the projects within the organization. Executives may require only an overview course of three to six hours, whereas employees who are more actively involved in the day-to-day activities of projects may require week-long training programs.

ROADBLOCKS

Training programs alone cannot overcome the fears and apprehensions that exist in the management ranks concerning the implementation of project management. Figure 5–2 illustrates the most common roadblocks that prevent an organization from completing Level 1.

• Resistance to Change
• Leaving Well Enough Alone
• Not Invented Here
• It Does Not Apply to Us
• We Don't Need It

FIGURE 5–2. Roadblocks to completion of Level 1.

Resistance to change is the result of management's belief that the implementation of project management will cause "culture shock," where functional managers will have to surrender some or all of their authority to the project managers. As a result, numerous excuses will appear as to why project management is not needed or will not work. Typical comments include: "We don't need it." "It doesn't apply to our business." "Let's leave well enough alone."

The implementation of project management does *not* have to be accompanied by shifts in the power and authority spectrum. However, there may be a shift in the reporting structure, inasmuch as project management is almost always accompanied by multiple-boss reporting. All training programs on project management emphasize multiple-boss reporting.

ADVANCEMENT CRITERIA

There are five key actions required before the organization can advance to Level 2. They are:

- Arrange for initial training and education in project management.
- Encourage the training (or hiring) of certified project management professionals (PMPs).
- Encourage employees to begin communicating in common project management language.

- Recognize available project management tools.
- Develop an understanding of the principles of project management: the project management body of knowledge as spelled out in the PMBOK® Guide.

The last item may prove the most difficult in non–project-driven organizations where project management is not regarded as a profession.

The successful completion of Level 1 usually occurs with a medium degree of difficulty. The time period to complete Level 1 could be measured in months or years, based upon such factors as:

- Type of company (project-driven versus non–project-driven)
- Size and nature of the projects
- Amount of executive support
- Visibility of executive support
- Strength of the existing corporate culture
- Previous experience, if any, with project management
- Corporate profitability
- Economic conditions (inflation, recession, etc.)
- The speed by which training can be accomplished

RISK

Level 1 carries a medium degree of risk. The organization might very well be resistant to change. Management may be fearful of a shift in the balance of power and authority.

Another major problem at Level 1 is when the organization first recognizes the complexities of multiple-boss reporting, which is a necessity for project management. Multiple-boss reporting can affect the wage and salary administration program and how employees are evaluated.

Typical factors that cause Level 1 to present a medium level of risk include:

- Fear of organizational restructuring
- Fear of changes in roles and responsibilities
- Fear of changes in priorities

ASSESSMENT INSTRUMENT FOR LEVEL 1

Completion of Level 1 is based upon gaining knowledge of the fundamental principles of project management and its associated terminology. The requirements for completing Level 1 can be fulfilled through a good understanding of the PMBOK® Guide prepared by the Project Management Institute (PMI).

Testing on the PMBOK® Guide is a good indicator of where you stand in relation to Level 1. The testing can be accomplished on an individual basis or by taking the average score from a group of individuals.

Below are 80 questions covering PMBOK® Guide and the basic principles of project management. There are five answers for each question. Although some of the answers may appear quite similar, you must select one and only one answer. After you finish Question 80, you will be provided with written instructions on how to grade the exercise.

QUESTIONS

1. A comprehensive definition of scope management would be:
   A. Managing a project in terms of its objectives through all life cycle phases and processes
   B. Approval of the scope baseline
   C. Approval of the detailed project charter
   D. The processes required to ensure that the project includes all the work required to complete the project successfully

2. The most common types of schedules include all but one of the following:
   A. Project network diagrams with date information added
   B. Resource leveling heuristics
   C. Bar charts
   D. Milestones

3. The communications environment involves both internal factors and external factors. An example of a typical internal factor is:
   A. Power games
   B. Business environment
   C. Technical state-of-the art
   D. Political environment

4. The most effective means of determining the cost of a project is to price out the:
   A. Work breakdown structure (WBS)
   B. Linear responsibility chart
   C. Project charter
   D. Scope statement

5. Employee unions would most likely satisfy which level in Maslow's hierarchy of needs?
   A. Social
   B. Self-Actualization
   C. Esteem
   D. Physiological

6. A document that describes the procurement item in sufficient detail to allow prospective sellers to determine if they can provide it is a:
   A. Contractual provision
   B. Statement of work (SOW)
   C. Terms and conditions statement
   D. Proposal

7. Future events or outcomes that are favorable are called:
   A. Risks
   B. Opportunities
   C. Surprises
   D. Contingencies

8. An example of an appraisal cost in terms of the cost of quality is:
   A. Surveys of vendors, suppliers, and subcontractors
   B. Evaluations of customer complaints
   C. Internal-external design reviews
   D. Process studies

9. Perhaps the biggest problem facing the project manager during integration activities within a matrix structure is:
   A. Coping with employees who report to multiple bosses
   B. Too much sponsorship involvement
   C. Unclear functional understanding of the technical requirements
   D. Escalating project costs

10. If you wish to compare actual project results to planned or expected results, you should:
    A. Hold a performance review
    B. Request a progress report
    C. Conduct a trend analysis
    D. Conduct a variance analysis

11. Communications has many different dimensions. Deciding to form a group among project managers in your organization to discuss lessons learned and best practices to follow is an example of which of the following dimensions:
    A. Internal
    B. External
    C. Horizontal
    D. Vertical

12. Which of the following methods is best suited to identifying the "vital few":
    A. Pareto analysis
    B. Cause-and-effect analysis
    C. Trend analysis
    D. Process control charts

13. A collection of formal procedures that includes the steps by which official project documents will be changed is defined through:
    A. The project management information system
    B. The change control system
    C. The Change Control Board
    D. Performance reports

14. A risk is noted by having a cause and:
    A. If it occurs, it only has a negative effect on the project's objectives
    B. A known unknown
    C. If it occurs, it has a consequence
    D. A constraint

15. In general, differences between and among project stakeholders should be resolved in favor of the:
   A. Project sponsor
   B. Performing organization
   C. Functional manager
   D. Customer

16. Project life cycles share many common characteristics, which include all of the following except:
   A. Increased ability for stakeholders to influence the final characteristics of the project toward the end of the life cycle
   B. Probability of successful completion being at the lowest at the beginning of the project
   C. Reduced ability of stakeholders to influence the final cost of the project as the project continues
   D. A low staffing level at the start of the project

17. Smoothing out resource requirements from period to period is called:
   A. Resource allocation
   B. Resource partitioning
   C. Resource leveling
   D. Resource quantification

18. The difference between the EV (Earned Value) and the PV (Planned Value) is referred to as:
   A. The schedule variance
   B. The cost variance
   C. The estimate of completion
   D. The actual cost of the work performed

19. Project managers must use a number of different interpersonal influences on projects to contribute to project success. If the project manager is viewed as being empowered to issue orders, he or she is using which of the following types of power:
   A. Expert
   B. Reward
   C. Referent
   D. Legitimate

20. The sender-receiver model in project communications includes:
   A. The choice of media
   B. The feedback loops and barriers to communications
   C. The presentation and meeting management techniques
   D. The choice of technology

21. A deliverable-oriented grouping of project components to organize and define the total project scope is:
   A. A detailed plan
   B. A linear responsibility chart
   C. A work breakdown structure (WBS)
   D. A cost accounting coding system

22. Modern quality management and project management are complementary as both disciplines recognize the importance of all but one of the following:
    A. Customer satisfaction
    B. Processes within phases
    C. Management responsibility
    D. Inspection over prevention

23. In which of the following circumstance(s) would you be most likely to buy goods or services instead of producing them in-house?
    A. Your company has excess capacity and can produce the goods or services
    B. Your company lacks capacity
    C. There are many reliable vendors for the goods or services that you are attempting to procure, but the vendors cannot achieve your level of quality
    D. Your company has an ongoing need for the item

24. A limitation of the bar chart is:
    A. Difficulty in changing it once it is prepared
    B. Hard to understand it if you do not have a knowledge of project management
    C. Difficulty in adding new items to it as the project changes
    D. Difficulty in performing any sensitivity analysis as it does not show the uncertainty involved in performing activities

25. The tool and technique used for risk management planning is:
    A. Assessment of stakeholder risk tolerances
    B. Planning meetings
    C. Documentation reviews
    D. Assumption and constraint analyses

26. Typically, during which phase in a project life cycle are most of the project expenses incurred:
    A. Concept phase
    B. Development or design phase
    C. Execution phase
    D. Termination phase

27. Going from Level 3 to Level 4 in the work breakdown structure (WBS) will result in:
    A. Less estimating accuracy
    B. Better control of the project
    C. Lower status reporting costs
    D. A greater likelihood that some key project element has been overlooked

28. Conflict management requires problem-solving. Which of the following is often referred to as a problem-solving technique and used extensively by project managers in conflict resolution:
    A. Confrontation
    B. Compromise
    C. Smoothing
    D. Forcing

29. Estimating the effect of the change of one project variable upon the overall project is known as:
    A. The project manager's risk aversion quotient
    B. The total project risk
    C. The expected value of the project
    D. Sensitivity analysis

30. Power games, withholding information, and hidden agendas are examples of:
    A. Feedback
    B. Communication barriers
    C. Indirect communication
    D. Mixed messages

31. The basic terminology for networks includes:
    A. Activities, events, personnel, skill levels, and slack
    B. Activities, documentation, events, personnel, and skill levels
    C. Slack, activities, events, and time estimates
    D. Time estimates, slack, sponsorship involvement, and activities

32. The "control points" in the work breakdown structure (WBS) used for assignments to specific organizational units or individuals are:
    A. Work packages
    B. Subtasks
    C. Tasks
    D. Code of accounts

33. Establishing a market window on a technology project or achieving a government-mandated compliance with environmental remediation are examples of:
    A. Imposed dates
    B. Weather restrictions on outdoor activities
    C. Major milestones
    D. Product characteristics

34. An example of a constraint to consider during procurement planning is:
    A. Indirect costs
    B. Funds availability
    C. Market conditions
    D. Procurement resources

35. The basic elements of a communication model include:
    A. Written and oral and listening and speaking
    B. Communicator, encoding, message, medium, decoding, receiver, and feedback
    C. Reports and briefings as well as memos and ad hoc conversations
    D. Reading, writing, participating in meetings, and listening

36. Assume that you are managing a project that is a joint venture between your company and two other firms. The project's quality policy then should be:
    A. Your responsibility to prepare
    B. The same as that of your customer
    C. The same as that of your company
    D. Prepared by the project team

37. The three most common types of project cost estimates are:
   A. Order of magnitude, parametric, and budget
   B. Parametric, definitive, and top down
   C. Order of magnitude, definitive, and bottom up
   D. Order of magnitude, budget, and definitive

38. Good project objectives must be:
   A. General rather than specific
   B. Established without considering resource constraints
   C. Realistic and attainable
   D. Measurable, intangible, and verifiable

39. The process of determining which risks might affect the project and documenting their characteristics is:
   A. Risk identification
   B. Risk response planning
   C. Risk management planning
   D. Qualitative risk analysis

40. In which type of contract arrangement is the *contractor* most likely to control costs?
   A. Cost-plus-fixed fee
   B. Firm-fixed price
   C. Time and materials
   D. Fixed-price-incentive firm target

41. A project can best be defined as:
   A. A series of nonrelated activities designed to accomplish single or multiple objectives
   B. A coordinated effort of related activities designed to accomplish a goal without a well-established end point
   C. Cradle-to-grave activities that must be accomplished in less than one year and consume human and nonhuman resources
   D. Any undertaking with a definable time frame and well-defined objectives that consumes both human and nonhuman resources with certain constraints

42. Risk management decision-making falls into three broad categories:
   A. Certainty, risk, and uncertainty
   B. Probability, risk, and uncertainty
   C. Probability, risk event, and uncertainty
   D. Hazard, risk event, and uncertainty

43. A process is considered to be out of control when there are which of the following consecutive data points (minimum) on either side of the mean on a control chart:
   A. 3
   B. 7
   C. 9
   D. 11

44. The work breakdown structure (WBS), the work packages, and the company's accounting system are tied together through:
   A. The code of accounts
   B. The overhead rates
   C. The budgeting system
   D. The capital budgeting process

45. A program can best be described as:
   A. A grouping of related activities that lasts two years or more
   B. A collection of projects and other work designed to meet strategic business objectives
   C. A group of projects managed in a coordinated way to obtain benefits not available from managing them individually
   D. A product line

46. Which of the following types of power comes through the organizational hierarchy:
   A. Coercive, legitimate, referent
   B. Reward, coercive, expert
   C. Referent, expert, legitimate
   D. Legitimate, coercive, reward

47. The most common definition of project success is:
   A. Within time
   B. Within time and cost
   C. Within time, cost, and technical performance requirements
   D. Within time, cost, performance, and acceptance by the customer/user

48. Activities with zero time duration are referred to as:
   A. Critical path activities
   B. Non–critical path activities
   C. Slack time activities
   D. Dummies

49. The procurement planning process should be accomplished during:
   A. Scope definition
   B. Solicitation planning
   C. Project initiation
   D. Scope planning

50. Project cash reserves are often used for adjustments in escalation factors, which may be beyond the control of the project manager. Other than possible financing (interest) cost and taxes, the three most common escalation factors involve changes in:
   A. Overhead rates, labor rates, and material costs
   B. Rework, cost-of-living adjustments, and overtime
   C. Material costs, shipping costs, and scope changes
   D. Labor rates, material costs, and cost reporting

51. The critical path in a network is the path that:
   A. Has the greatest degree of risk
   B. Is the longest during the project and determines its duration
   C. Must be completed before all other paths
   D. Has activities with float greater than zero

52. The major difference between project and line management is that the project manager may not have any control over which basic management function:
    A. Decision-making
    B. Staffing
    C. Tracking/monitoring
    D. Reviewing

53. During which phase of a project is the uncertainty the greatest:
    A. Design
    B. Development/execution
    C. Concept
    D. Closing

54. Quality often is confused with grade. This means that:
    A. Low quality is always a problem, but low grade may not be a problem
    B. Low grade is always a problem, along with low quality
    C. Quality is defined as a category or rank with entities having the same functional use but different technical characteristics
    D. Grade is defined as the total characteristics of an entity that bear on its ability to satisfy stated or implied needs

55. Project managers need exceptionally good communication and negotiation skills primarily because:
    A. They may be leading a team over which they have no direct control
    B. This need is mandated by the project's procurement activities
    C. They are expected to be technical experts
    D. They must provide executive/customer/sponsor briefings

56. For effective communication, the message should be oriented to:
    A. The initiator
    B. The receiver
    C. The management style
    D. The corporate culture

57. Common factors that may constrain how the project team is organized include all but one of the following:
    A. The structure of the performing organization
    B. Preferences of the team
    C. Expected staff assignments
    D. Responsibility Assignment Matrix

58. On a precedence diagram, the arrow between two boxes is called:
    A. An activity
    B. A constraint
    C. An event
    D. The critical path

59. In which type of contract arrangement is the *contractor* least likely to control costs:
    A. Cost-plus-incentive fee
    B. Firm-fixed price
    C. Fixed-price-award fee
    D. Purchase order

60. The financial closeout of a project dictates that:
- **A.** All project funds have been spent
- **B.** No charge numbers have been overrun
- **C.** No follow-on work from this client is possible
- **D.** No further charges can be made against the project

61. A graphical display of accumulated costs and labor hours for both budgeted and actual costs, plotted against time, is called:
- **A.** A trend line
- **B.** A trend analysis
- **C.** An S curve
- **D.** A percent completion report

62. If you are using a control chart and find that the process is in control, it is important to recognize that:
- **A.** The process should not be adjusted
- **B.** The process should not be changed to provide improvements
- **C.** Sources of random variation can be easily changed without the need to restructure the process
- **D.** Sources of random variation are never present

63. The major difference between PERT and CPM networks is:
- **A.** PERT requires three time estimates whereas CPM uses one time estimate
- **B.** PERT is used only for construction projects whereas CPM is used solely for R&D
- **C.** PERT addresses only time whereas CPM also includes costs and resource availability
- **D.** PERT is measured in days whereas CPM uses weeks or months

64. Information can be shared by team members and other stakeholders using a variety of information retrieval systems including:
- **A.** Project meetings
- **B.** Fax
- **C.** Electronic mail
- **D.** Electronic databases

65. Assume that you have decided to use mitigation as a risk response technique. This means that:
- **A.** You are shifting consequences of a risk to another party
- **B.** You are reducing the probability and/or consequences of an adverse risk event to an acceptable threshold
- **C.** You now need to establish a contingency allowance
- **D.** Your next step should be to prepare a fallback plan

66. The traditional or functional organizational form has the disadvantage of:
- **A.** Poorly established vertical communications channels
- **B.** No single focal point for clients/sponsors
- **C.** Ineffective technical control
- **D.** Inflexible use of personnel

67. Which of the following is not a basic element of contracts:
- **A.** Consideration
- **B.** Mutual agreement
- **C.** Level of effort
- **D.** Legal purpose

68. Taking action to increase the effectiveness and efficiency of the project to provide added benefits to the stakeholders is the purpose of:
   A. Quality planning
   B. Inspections
   C. Quality audits
   D. Quality improvement

69. During the procurement planning process, it is important to assess the current project boundaries. This can be done by reviewing the:
   A. Results of the make-or-buy analysis
   B. Product description
   C. Scope statement
   D. Constraints and assumptions

70. In project communications management, in order to ensure that the information needs of various stakeholders are met, you should:
   A. Prepare a stakeholder analysis
   B. Establish an information distribution system
   C. Assess communications skills
   D. Evaluate available communications technologies

71. Assigning resources in an attempt to find the shortest project schedule consistent with *fixed* resource limits is called:
   A. Resource allocation
   B. Resource partitioning
   C. Resource leveling
   D. Resource quantification

72. The process of assessing the impact and likelihood of identified risks is known as:
   A. Risk management planning
   B. Risk response planning
   C. Qualitative risk analysis
   D. Quantitative risk analysis

73. An advantage of the analogous cost-estimating technique is:
   A. Greater accuracy from its use
   B. Historical information is not required
   C. Expert judgment is never needed
   D. Lower costs are involved in its use

74. Action taken to bring a defective or nonconforming item in compliance with requirements or specifications is the purpose of:
   A. Rework
   B. Control charts
   C. Audits
   D. Process adjustments

75. If you want to describe where the project now stands, you should:
   A. Prepare an estimate to complete
   B. Prepare an earned value analysis
   C. Prepare a status report
   D. Prepare a progress report

76. One purpose of risk control is to:
    A. See if assumptions are still valid
    B. Determine whether risk response actions are as effective as expected
    C. Assess whether a risk trigger has occurred
    D. Take corrective action

77. In source selection a weighting system may be used for all but one of the following purposes:
    A. To rank order all proposals to establish a negotiating sequence
    B. To select a single source who will be asked to sign a standard contract
    C. To establish minimum requirements of performance for one or more evaluation criteria
    D. To quantify qualitative data to minimize the effect of personal prejudice on source selection

78. The overall intentions and directions of an organization with regard to quality is the purpose of:
    A. The total quality management movement
    B. The quality assurance process
    C. The quality planning process
    D. The organization's quality policy

79. The project communications management plan should:
    A. State communications skills to use
    B. Describe methods used to gather and store information
    C. Provide information to stakeholders as to how resources are being used to meet project objectives
    D. Describe relationships between the organization and stakeholders

80. During a project review meeting, we discover that the planned value is $400,000, the actual costs are $325,000, and the earned value is $300,000. We can therefore conclude that:
    A. The project is behind schedule and overrunning costs
    B. The project is ahead of schedule, but costs are higher than budgeted
    C. The project is behind schedule with costs under control
    D. The project is on schedule, but costs are higher than budgeted

Using the answer key, score yourself and fill in the tables in Exhibit 1. Give yourself 10 points for each correct answer and no points for an incorrect answer. After you fill in the tables in Exhibit 1, continue on for an interpretation of your results.

ANSWER KEY

1. D	6. B
2. B	7. B
3. A	8. C
4. A	9. A
5. A	10. D

11. C	46. D
12. A	47. D
13. B	48. D
14. C	49. A
15. D	50. A
16. A	51. B
17. C	52. B
18. A	53. C
19. D	54. A
20. B	55. A
21. C	56. B
22. D	57. D
23. B	58. B
24. D	59. A
25. B	60. D
26. C	61. C
27. B	62. A
28. A	63. A
29. D	64. D
30. B	65. B
31. C	66. B
32. A	67. C
33. A	68. D
34. B	69. C
35. B	70. A
36. D	71. A
37. D	72. C
38. C	73. D
39. A	74. A
40. B	75. C
41. D	76. D
42. A	77. C
43. B	78. D
44. A	79. B
45. C	80. A

Exhibit 1

Put the points in the space provided by each question and then total each category.

Scope Management	Time Management	Cost Management
1. _____	2. _____	4. _____
16. _____	17. _____	10. _____
21. _____	24. _____	18. _____
27. _____	31. _____	26. _____
32. _____	33. _____	37. _____
38. _____	48. _____	44. _____
41. _____	51. _____	50. _____
45. _____	58. _____	61. _____
47. _____	63. _____	73. _____
60. _____	71. _____	80. _____
TOTAL _____	TOTAL _____	TOTAL _____

Human Resources Management	Procurement Management	Quality Management
5. _____	6. _____	8. _____
9. _____	13. _____	12. _____
15. _____	23. _____	22. _____
19. _____	34. _____	36. _____
28. _____	40. _____	43. _____
46. _____	49. _____	54. _____
52. _____	59. _____	62. _____
55. _____	67. _____	68. _____
57. _____	69. _____	74. _____
66. _____	77. _____	78. _____
TOTAL _____	TOTAL _____	TOTAL _____

Risk Management	Communications Management
7. _____	3. _____
14. _____	11. _____
25. _____	20. _____
29. _____	30. _____
39. _____	35. _____
42. _____	56. _____

53. _____		64. _____	
65. _____		70. _____	
72. _____		75. _____	
76. _____		79. _____	
TOTAL _____		TOTAL _____	

Category	Points
Scope Management:	_____
Time Management:	_____
Cost Management:	_____
Human Resources Management:	_____
Procurement Management:	_____
Quality Management:	_____
Risk Management:	_____
Communications Management:	_____
Total:	_____

EXPLANATION OF POINTS FOR LEVEL 1 _____

If you received a score of 60 or more points in each of the eight categories, then you have a reasonable knowledge of the basic principles of project management. If you received a score of 60 or more in all but one or two of the categories, it's possible that you and your organization still possess all the knowledge you need of basic principles but that one or two of the categories do not apply directly to your circumstances. For example, if most of your projects are internal to your organization, procurement management may not be applicable. Also, for internal projects, companies often do not need the rigorous cost control systems that would be found in project-driven organizations. Eventually, however, specialized training in these deficient areas will be needed.

If your score is less than 60 in any category, a deficiency exists. For scores less than 30 in any category, rigorous training programs on basic principles appear necessary. The organization appears highly immature in project management.

A total score on all categories of 600 or more would indicate that the organization appears well positioned to begin work on Level 2 of the PMMM. If your organization as a whole scores less than 600 points, there may exist pockets of project management. Each pocket may be at a different level of knowledge. Project-driven pockets generally possess more project management knowledge than non–project-driven pockets.

This assessment instrument can be used to measure either an individual's knowledge or an organization's knowledge. To assess organizational knowledge accurately, however, care must be taken in determining the proper cross-section of participants to be tested.

Level 2: Common Processes

6

INTRODUCTION

Learning the basics of project management, and even having several employees certified as Project Management Professionals (PMPs), does not guarantee that project management is being used in your organization. Even if it is being used, it may not be used effectively. Level 2 is the stage where an organization makes a concerted effort to use project management and to develop processes and methodologies to support its effective use.

In Level 2, the organization realizes that common methodologies and processes are needed such that managerial success on one project can be repeated on other projects. Also apparent in this level is the fact that certain behavioral expectations of organizational personnel are necessary for the repetitive execution of the methodology. These are the characteristics of Level 2, as shown in Figure 6–1:

- Tangible benefits of using project management must become apparent. The most common benefits include lower cost, shortened schedules, no sacrifice of scope or quality, and the potential for a higher degree of customer satisfaction.
- Project management must be supported throughout all levels of the organization, including the senior levels. It is possible that changes to the corporate culture may be necessary, thus mandating executive support.
- A continuous stream of successfully managed projects requires methodologies and processes that can be used over and over again. This requires an organizational commitment.

Common Processes
• Recognition of Benefits of Project Management
• Organizational Support at All Levels
• Recognition of Need for Processes/Methodologies
• Recognition of the Need for Cost Control
• Development of a Project Management Training Curriculum

FIGURE 6–1. Characteristics of Level 2.

- Managing projects within scope and time is only part of the effort. The projects must also be completed within cost, and this may mandate changes to the cost accounting system.
- The final characteristic of Level 2 is the development of a project management curriculum rather than just a project management course. This is often seen as proof of the organization's firm commitment to project management.

These bullets are the outputs of the common processes. In other words, do I have common processes to facilitate repeatability? It should be noted that many of the benefits of common processes many not be clearly visible. The benefits may be intangible at first, and then become visible benefits later.

LIFE CYCLES FOR LEVEL 2

Common processes require a good process definition effort accompanied by the necessary organizational behavior needed for the execution of the processes. Level 2, common processes, can be broken down into five life cycle phases, as shown in Figure 6–2. These life cycle phases are actually subphases or steps within Level 2 of the project management maturity model (PMMM) in order to develop common processes. The first life cycle phase of Level 2 is the embryonic phase, which is where the organization recognizes that project management can benefit the organization. The embryonic phase includes:

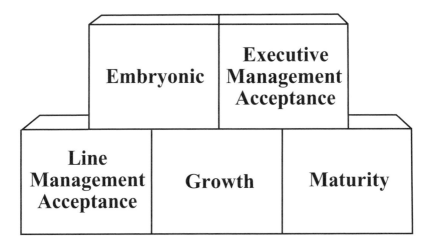

FIGURE 6–2. Life cycle phases for Level 2 of project management maturity.

- Recognizing the need for project management
- Recognizing the potential benefits of project management
- Recognizing the applications of project management to the various parts of the business
- Recognizing some of the changes necessary to implement project management

Companies do not generally promote the acceptance of project management unless they understand a sound basis for wanting project management. The six most common driving forces for project management are as follows:

- Capital projects: High-dollar-value capital projects require effective planning and scheduling. Without project management, ineffective use of manufacturing resources may occur.
- Customer expectations: Customers have the right to expect the contractor to manage the customer's work requirements efficiently and effectively.
- Internal competitiveness: Executives want employees to focus on external competition rather than internal competition, power struggles, and gamesmanship.
- Executive understanding: Although it's uncommon, executives can drive the acceptance of project management from the top of the organization down to the bottom.
- New product development: Executives want a methodology in place that provides a high likelihood that R&D projects will be completed successfully, in a timely manner, and within reasonable cost.
- Efficiency and effectiveness: Executives want the organization to be highly competitive.

In theory, most companies have one and only one driving force. While we've just discussed six different driving forces, in practice, in reality, they combine to give us one, and only one—survival. This is shown in Figure 6–3. Once executives recognize that project management is needed for survival, changes occur quickly.

What is unfortunate about the embryonic phase is that the recognition of benefits and applications may be seen first by lower and middle levels of management. Senior management must then be "sold" on the concept of project management. This leads us to the second life cycle phase, executive management acceptance. Included in the executive management acceptance phase are the following:

- Visible executive support
- Executive understanding of project management
- Project sponsorship
- Willingness to change the way the company does business

The third life cycle phase of Level 2 is line management acceptance. This includes:

- Visible line management support
- Line management commitment to project management
- Line management education
- Release of functional employees for project management training programs

It is highly unlikely that line managers will provide support for project management unless they also see "visible" executive support.

The fourth life cycle phase of Level 2 is the growth phase. This is the critical phase. Although some of the effort in this phase can be accomplished in parallel with the first three life cycle phases of Level 2, the completion of this phase

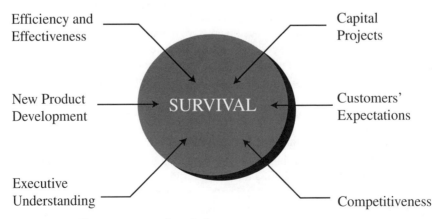

FIGURE 6–3. The components of survival.

is predicated upon the completion of the first three life cycle phases. The growth phase is the beginning of the creation of the project management process. Included in this phase are:

- Development of company project management life cycles
- Development of a project management methodology
- A commitment to effective planning
- Minimization of scope changes (i.e., of creeping scope)
- Selection of project management software to support the methodology

Unfortunately, companies often develop several types of methodologies for each type of project within the organization. This becomes an inefficient use of resources, although it can function as a good learning experience for the company.

The fifth life cycle phase of Level 2 is the so-called "initial maturity phase" of Level 2. Included in this phase are:

- The development of a management cost/schedule control system
- Integration of schedule and cost control
- Development of an ongoing educational curriculum to support project management and enhance individual skills

Many companies never fully complete this life cycle phase because the organization is resistant to project cost control, otherwise known as horizontal accounting. Line managers dislike horizontal accounting because it clearly identifies which line managers provide good estimates for projects and which do not. Executives resist horizontal accounting because the executives want to establish a budget and schedule long before a project plan is created.

ROADBLOCKS

Figure 6–4 illustrates the most common roadblocks that prevent an organization from completing Level 2. Based upon the strength and longevity of the corporate culture, there could be strong resistance to change. The argument is always, "What we already have works well." The resistance to change stems from the fear that support for a new methodology will result in a shift in the established power and authority relationships.

Another area of resistance is due to the misbelief that a new methodology *must* be accompanied by rigid policies and procedures, thus once again causing potential changes to the power and authority structure. The final roadblock comes from the fear that "horizontal accounting" will bring to the surface problems that people would prefer to keep hidden, such as poor estimating ability.

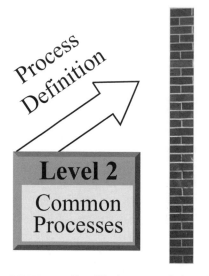

Process Definition

Level 2

Common Processes

- Resistance to a New Methodology
- What We Already Have Works Well
- Believing That a Methodology Needs Rigid Policies and Procedures
- Resistance to "Horizontal" Accounting

FIGURE 6–4.　Roadblocks to completion of Level 2.

ADVANCEMENT CRITERIA

There are four key actions required to complete Level 2 and advance to Level 3. These actions are as follows:

- Develop a culture that supports both the behavioral and quantitative sides of project management.
- Recognize both the driving forces/need for project management and the benefits that can be achieved in both the short term and the long term.
- Develop a project management process/methodology such that the desired benefits can be achieved on a repetitive basis.
- Develop an ongoing, all-employee project management curriculum such that the project management benefits can be sustained and improved upon for the long term.

RISK

The successful completion of Level 2 usually occurs with a medium degree of difficulty. The time period to complete Level 2 is usually six months to two years, based upon such factors as:

- Type of company (project-driven versus non–project-driven)
- Visibility of executive support
- Strength of the corporate culture

- Resistance to change
- Speed with which a good, workable methodology can be developed
- Existence of an executive-level champion to drive the development of the project management methodology
- Speed with which the project management benefits can be realized

The risk in this level can be overcome through strong, visible executive support.

OVERLAPPING LEVELS

Level 2 can and does overlap Level 1. There is no reason why we must wait for a multitude of people to be trained in project management before we begin the development of processes and methodologies. Also, the earlier the company begins developing processes and methodologies, the earlier those processes and methodologies can be included as part of the training. One HMO conducted a three-day course on the principles of project management. A fourth day was spent covering the company's processes and methodologies for project management. Thus the employees could see clearly how the processes/methodologies utilized the basic concepts of project management.

ASSESSMENT INSTRUMENT FOR LEVEL 2

Level 2, common processes, is the process definition level. Level 2 can be fulfilled by recognizing the different life cycle phases of Level 2.

The following 20 questions explore how mature you believe your organization to be in regard to Level 2 and the accompanying life cycle phases of Level 2. Beside each question you will circle the number that corresponds to your opinion. In the example below, your choice would have been "Slightly Agree."

$$
\begin{array}{rl}
-3 & \text{Strongly Disagree} \\
-2 & \text{Disagree} \\
-1 & \text{Slightly Disagree} \\
0 & \text{No Opinion} \\
\boxed{+1} & \text{Slightly Agree} \\
+2 & \text{Agree} \\
+3 & \text{Strongly Agree}
\end{array}
$$

Example: $(-3, \quad -2, \quad -1, \quad 0, \quad \boxed{+1}, \quad +2, \quad +3)$

The row of numbers from -3 to $+3$ will be used later for evaluating the results. After answering Question 20, you will grade the exercise.

QUESTIONS

The following 20 questions involve Level 2 *maturity*. The questions look at both repeatability of processes and actions necessary to support continuous development of processes. Please answer each question as honestly as possible. Circle the answer you feel is correct.

1. My company recognizes the *need* for project management. This *need* is recognized at all levels of management, including senior management. $(-3 \quad -2 \quad -1 \quad 0 \quad +1 \quad +2 \quad +3)$

2. My company has a system in place to manage both cost and schedule. The system requires charge numbers and cost account codes. The system *reports variances* from planned targets. $(-3 \quad -2 \quad -1 \quad 0 \quad +1 \quad +2 \quad +3)$

3. My company has recognized the *benefits* that are possible from implementing project management. These *benefits* have been recognized at all levels of management, including senior management. $(-3 \quad -2 \quad -1 \quad 0 \quad +1 \quad +2 \quad +3)$

4. My company (or division) has a well-definable project management methodology using life cycle phases. $(-3 \quad -2 \quad -1 \quad 0 \quad +1 \quad +2 \quad +3)$

5. Our executives visibly support project management through executive presentations, correspondence, and by occasionally attending project team meetings/briefings. $(-3 \quad -2 \quad -1 \quad 0 \quad +1 \quad +2 \quad +3)$

6. My company is committed to quality up-front planning. We try to do the best we can at planning. $(-3 \quad -2 \quad -1 \quad 0 \quad +1 \quad +2 \quad +3)$

7. Our lower- and middle-level line managers totally and visibly support the project management process. $(-3 \quad -2 \quad -1 \quad 0 \quad +1 \quad +2 \quad +3)$

8. My company is doing everything possible to minimize "creeping" scope (i.e., scope changes) on our projects. $(-3 \quad -2 \quad -1 \quad 0 \quad +1 \quad +2 \quad +3)$

9. Our line managers are committed not only to project management, but also to the promises made to project managers for deliverables. $(-3 \quad -2 \quad -1 \quad 0 \quad +1 \quad +2 \quad +3)$

10. The executives in my organization have a good understanding of the principles of project management. $(-3 \quad -2 \quad -1 \quad 0 \quad +1 \quad +2 \quad +3)$

11. My company has selected one or more
    project management software packages
    to be used as the project tracking
    system. $(-3 \quad -2 \quad -1 \quad 0 \quad +1 \quad +2 \quad +3)$

12. Our lower- and middle-level line
    managers have been trained and
    educated in project management. $(-3 \quad -2 \quad -1 \quad 0 \quad +1 \quad +2 \quad +3)$

13. Our executives both understand project
    sponsorship and serve as project
    sponsors on selected projects. $(-3 \quad -2 \quad -1 \quad 0 \quad +1 \quad +2 \quad +3)$

14. Our executives have recognized or
    identified the *applications* of project
    management to various parts of our
    business. $(-3 \quad -2 \quad -1 \quad 0 \quad +1 \quad +2 \quad +3)$

15. My company has successfully *integrated*
    cost and schedule control for both
    managing projects and reporting status. $(-3 \quad -2 \quad -1 \quad 0 \quad +1 \quad +2 \quad +3)$

16. My company has developed a project
    management curriculum (i.e., more than
    one or two courses) to enhance the
    project management skills of our
    employees. $(-3 \quad -2 \quad -1 \quad 0 \quad +1 \quad +2 \quad +3)$

17. Our executives have recognized what
    must be done in order to achieve
    maturity in project management. $(-3 \quad -2 \quad -1 \quad 0 \quad +1 \quad +2 \quad +3)$

18. My company views and treats project
    management as a profession rather than
    a part-time assignment. $(-3 \quad -2 \quad -1 \quad 0 \quad +1 \quad +2 \quad +3)$

19. Our lower- and middle-level line
    managers are willing to release their
    employees for project management
    training. $(-3 \quad -2 \quad -1 \quad 0 \quad +1 \quad +2 \quad +3)$

20. Our executives have demonstrated a
    willingness to change our way of doing
    business in order to mature in project
    management. $(-3 \quad -2 \quad -1 \quad 0 \quad +1 \quad +2 \quad +3)$

Now turn to Exhibit 2 and grade your answers.

Exhibit 2

Each response you circled in Questions 1–20 had a column value between -3 and $+3$. In the appropriate spaces below, place the circled value (between -3 and $+3$) beside each question.

Embryonic	**Executive**	**Line Management**
1. _____	5. _____	7. _____
3. _____	10. _____	9. _____
14. _____	13. _____	12. _____
17. _____	20. _____	19. _____
TOTAL _____	TOTAL _____	TOTAL _____

Growth	**Maturity**
4. _____	2. _____
6. _____	15. _____
8. _____	16. _____
11. _____	18. _____
TOTAL _____	TOTAL _____

Transpose your total score in each category to the table below by placing an "X" in the appropriate area.

Points													
Life Cycle Phases	**−12**	**−10**	**−8**	**−6**	**−4**	**−2**	**0**	**+2**	**+4**	**+6**	**+8**	**+10**	**+12**
Maturity													
Growth													
Line Management													
Executive													
Embryonic													

EXPLANATION OF POINTS FOR LEVEL 2

High scores (usually +6 or greater) for a life cycle phase indicate that these evolutionary phases of early maturity have been achieved or at least you are now in this phase. Phases with very low numbers have not been achieved yet.

Consider the following scores:

Embryonic:	+ 8
Executive:	+10
Line management:	+ 8
Growth:	+ 3
Maturity:	− 4

This result indicates that you have probably completed the first three stages and are now entering the growth phase. Keep in mind that the answers are not always this simple because companies can achieve portions of one stage in parallel with portions of a second or third phase.

Level 3: Singular Methodology

7

INTRODUCTION

Level 3 is the level in which the organization recognizes that synergism and process control can best be achieved through the development of a singular methodology rather than by using multiple methodologies for the same group or domain of projects. There may exist a separate methodology for new product development and another methodology for information systems. The goal, however, should be to determine the minimum number of domains or groups and have one methodology for each. In this level, the organization is totally committed to the concept of project management. The characteristics of Level 3, as shown in Figure 7–1, are as follows:

- Integrated processes: This is where the organization recognizes that multiple processes can be streamlined into one, integrated process encompassing all other processes. (However, not all companies have the luxury of using a single methodology.)
- Cultural support: Integrated processes create a singular methodology. It is through this singular methodology that exceptional benefits are achieved. The execution of the methodology is through the corporate culture, which now wholeheartedly supports the project management approach. The culture becomes a cooperative culture.
- Management support: In this level, project management support permeates the organization throughout all layers of management. The support is visible. Each layer or level of management understands its role and the support needed to make the singular methodology work.

Singular Methodology

- Integrated Processes
- Cultural Support
- Management Support at All Levels
- Informal Project Management
- Return on Investment for Project Management Training Dollars
- Behavioral Excellence

FIGURE 7–1. Characteristics of Level 3.

- Informal project management: With management support and a cooperative culture, the singular methodology is based upon guidelines and checklists, rather than based on the expensive development of rigid policies and procedures. Paperwork is minimized.
- Training and education: With strong cultural support, the organization realizes financial benefits from project management training. The benefits can be described quantitatively and qualitatively.
- Behavioral excellence: The organization recognizes the behavioral differences between project management and line management. Behavioral training programs are developed to enhance project management skills.

These six characteristics formulate the "hexagon of excellence," as shown in Figure 7–2. These six areas differentiate those companies excellent in project management from those with average skills in project management. Each of the six areas is discussed below.

INTEGRATED PROCESSES

Companies that are relatively immature in project management have multiple processes in place. Figure 7–3 shows the three most common of these separate processes. Concurrent engineering, for those unfamiliar with the term, is similar to fast-tracking a project where activities are overlapping in order to accelerate the completion date. Why, however, would a company want its processes, its facilities, its resources in general, to be totally uncoupled? The first two processes to be integrated, once an organization understands the advantages, are usually project management and total quality management (TQM). After all, employees trained in the principles of TQM will realize the similarities between the two

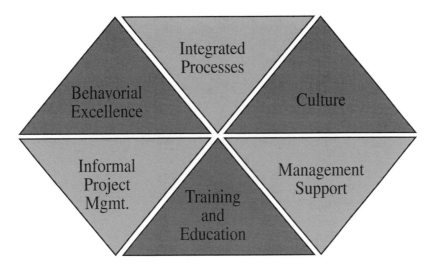

FIGURE 7–2. The hexagon of excellence.

processes. All of the winners of the prestigious Malcolm Baldrige Award have excellent project management systems in place.

When organizations begin to realize the importance of a singular methodology, project management becomes integrated with TQM and concurrent engineering to formulate a singular methodology. This integration is shown in Figure 7–4. As companies begin to climb the ladder toward excellence in project management, the initial singular methodology is further enhanced to include risk

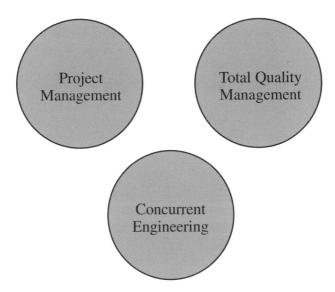

FIGURE 7–3. Totally uncoupled processes.

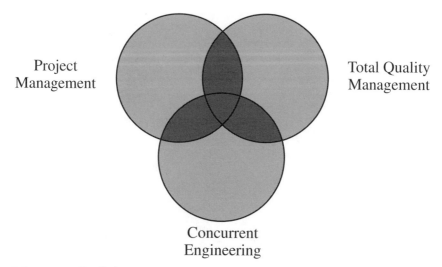

FIGURE 7–4. Totally integrated processes.

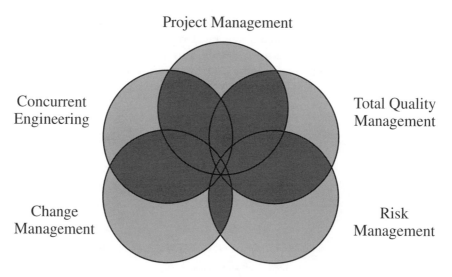

FIGURE 7–5. Integrated processes for the twenty-first century.

management and change management, as shown in Figure 7–5. Risks generally require scope changes, which, in turn, create additional risks. Creating a singular, integrated methodology that encompasses all other methodologies leads to organizational efficiency and effectiveness.

As a final note, not all companies have the luxury or ability to utilize one and only one methodology for the execution of all projects. Some companies readily admit that two methodologies might be needed: one for new product development and a second methodology for information systems projects. Therefore, it may be more appropriate to recognize this level as an attempt to get everyone to agree to a singular methodology, if possible.

CULTURE

Project management methodologies must not simply be pieces of paper. The pieces of paper must be converted into a world-class methodology by the way in which the corporate culture executes the methodology. Companies excellent in project management have cooperative cultures where the entire organization supports the singular methodology. Organizational resistance is at a minimum, and everyone pitches in during times of trouble.

Cultural transformation is never easy. From a project management perspective, common purposes for a cultural transformation include the corporate vision,

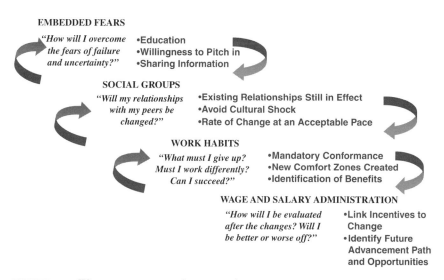

FIGURE 7–6. Ways to overcome resistance to change.

corporate goals, and the creation of a singular methodology for project management. Although there are a multitude of cultural issues, the four most common fears affecting project management are:

- The embedded fear of change
- The fear of having to create a new social group
- The fear of a change in work habits or comfort zone
- The uncertain impact on the wage and salary program

The hierarchy of fears can change from company to company, yet strategies must be put in place to overcome these fears or else Level 3 completion could be prolonged. Typical ways to overcome these fears are shown in Figure 7–6.

MANAGEMENT SUPPORT

Cooperative cultures require effective management support at all levels. During the execution of the project management methodology, the interface between project management and line management is critical. Effective relationships with line management are based upon these factors:

- Project managers and line managers share accountability for the successful completion of a project. Line managers must keep their promises to the project managers.
- Project managers negotiate with line managers for the accomplishment of deliverables rather than for specific talent. Project managers can request specific talent, but the final decision for staffing belongs to the line manager.
- Line managers trust their employees enough to empower those employees to make decisions related to their specific functional area without continuously having to run back to their line manager.
- If a line manager is unable to keep a promise he or she made to a project, then the project manager must do everything possible to help the line manager develop alternative plans.

The relationship between project management and senior management is equally important. A good relationship with executive management, specifically the executive sponsor, includes these factors:

- The project manager is empowered to make project-related decisions. This is done through decentralization of authority and decision-making.
- The sponsor is briefed periodically while maintaining a hands-off, but available, position.

- The project manager (and other project personnel) are encouraged to present recommendations and alternatives rather than just problems.
- Exactly what needs to be included in a meaningful executive status report has been formulated.
- A policy is in place calling for periodic, but not excessively frequent, briefings.

INFORMAL PROJECT MANAGEMENT

With informal project management, the organization recognizes the high cost of paperwork. Informal project management does not *eliminate* paperwork. Instead, paperwork requirements are *reduced* to the minimum acceptable levels. For this to work effectively, the organization must experience effective communications, cooperation, trust, and teamwork. These four elements are critical components of a cooperative culture.

As trust develops, project sponsorship may be pushed down from the executive levels to middle management. The project managers no longer wears multiple hats (i.e., being a project manager and line manager at the same time), but functions as a dedicated project manager.

The development of project management methodologies at Level 2 are based upon rigid policies and procedures. But in Level 3, with a singular methodology based more upon informal project management, methodologies are written in the format of general guidelines and checklists. This drastically lowers methodology execution cost and execution time.

The advent of colored printers has allowed companies to provide additional information without words. As an example, one company uses a "traffic light" beside each work breakdown structure (WBS) work package in the status report. The traffic light is either red, yellow, or green, based upon the following definitions:

- Red: A problem exists that may affect time, cost, scope, or quality. Sponsor involvement is necessary.
- Yellow: This is a caution. A potential problem may exist. The sponsor is informed, but no action by the sponsor is necessary right now.
- Green: Work is progressing as planned. Sponsor involvement is not necessary.

TRAINING AND EDUCATION

In Level 3, there is a recognition that there exists a return on investment for training dollars. The benefits, or return on investment, can be measured quantitatively and qualitatively. Quantitative results include:

- Shorter product development time
- Faster, higher quality decisions
- Lower costs
- Higher profit margins
- Fewer people needed
- Reduction in paperwork
- Improved quality and reliability
- Lower turnover of personnel
- Quicker "best practices" implementation

Qualitative results include:

- Better visibility and focus on results
- Better coordination
- Higher morale
- Accelerated development of managers
- Better control
- Better customer relations
- Better support from the functional areas
- Fewer conflicts requiring senior management involvement

Project management training and education is an investment and, as such, senior management wishes to know when the added profits will materialize. This can best be explained from Figure 7–7. Initially, there may be a substantial cost incurred during Level 2 and the beginning of Level 3. But as the culture develops

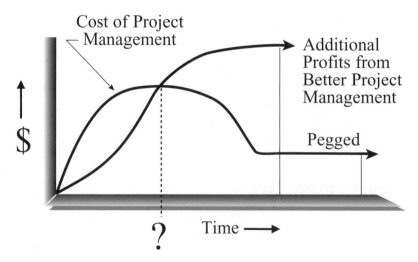

FIGURE 7–7. Project management costs versus benefits.

and informal project management matures, the cost of project management diminishes to a pegged level while the additional profits grow. The "question mark" in Figure 7–7 generally occurs during Level 3, which is usually about two to five years after the organization has made a firm commitment to project management.

A question normally asked by executives is, "How do we know if we are in Level 3 of the project management maturity model (the PMMM)?" The answer is by the number of conflicts coming up to the senior levels of management for resolution.

By Level 3, executives have realized that the speed by which the benefits can be achieved can be accelerated through proper training and education. Therefore the training and education in Level 3 does not consist merely of a few random courses. Instead, as discussed in the advancement criteria for completing Level 2 and moving up to Level 3, the company develops a project management curriculum. This will encompass a "core competency model" for the basic and advanced skills that a project manager should possess. Training is conducted to support the core competency skills.

BEHAVIORAL EXCELLENCE

Behavioral excellence occurs when the organization recognizes the differences between project management and line management, and the fact that a completely different set of training courses is required to support sustained project management growth. Emphasis is placed on:

- Motivation in project management
- Creation of outstanding project leaders
- Characteristics of productive teams
- Characteristics of productive organizations
- Sound and effective project management

People are often under the misapprehension that achieving Level 3 in the PMMM will deliver 100 percent successful projects. This is not true. Successful implementation of project management does *not* guarantee that your projects will be successful. Instead, it does guarantee that your projects will be managed effectively, thus improving your chances of success. From Figure 7–8 we see that, during Level 3 of the PMMM, the number of project successes increases. However, even though the number of successes increases, and comes to dramatically exceed the number of failures, failure still exists. Project management does not circumvent the problem of unrealistic objectives or targets, unforeseen acts of God, and economic upheaval. Any company that has a 100 percent project success rate is not working on enough projects. No risk is being taken. Also, any executive sponsor or project manager who always makes the right decision is not making enough decisions.

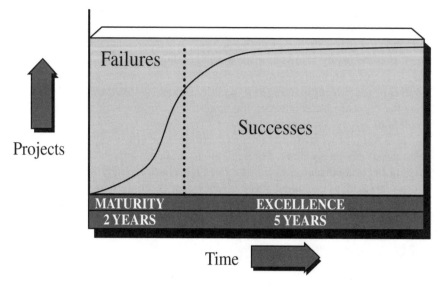

FIGURE 7–8. Growth in successes.

ROADBLOCKS

Figure 7–9 shows the key roadblocks that prevent an organization from completing Level 3. They include:

- Don't fix it if it isn't broken. We should continue to use the processes now in place.
- There will always exist initial resistance to a singular methodology for fear that it will be accompanied by shifts in the balance of power.
- Line managers may resist accepting accountability for the promises made to the projects. Shared accountability is often viewed as a high risk for the line managers.
- Organizations with strong, fragmented corporate cultures often resist being converted over to a single, cooperative culture.
- Some organizations thrive on the belief that what is not on paper has not been said. Overemphasis on documentation is a bad habit that is hard to break.

ADVANCEMENT CRITERIA

There are certain key actions needed to advance from Level 3 to Level 4. These key actions are as follows:

- Don't Fix It If It Isn't Broken
- Resistance to a Singular Integrated Methodology (i.e., Repeatable Process)
- Resistance to Shared Accountability
- Fragmented Corporate Culture
- Overemphasis on Documentation

FIGURE 7–9. Roadblocks to completion of Level 3.

- Integrate all related processes into a single methodology with demonstrated successful execution.
- Encourage the corporate-wide acceptance of a culture that supports informal project management and multiple-boss reporting.
- Develop support for shared accountability.

RISK

The successful completion of Level 3 is accompanied by a high degree of difficulty if the culture of the company must change. Culture shock may result. The time period to complete Level 3 is measured in years, based upon such factors as:

- The speed at which the culture will change
- The acceptance of informal project management
- The acceptance of a singular methodology

The greatest degree of risk in project management is attributed to the corporate culture. Poorly designed methodologies can convert a good, cooperative culture into a combative culture.

If an organization develops a singular methodology (or, if necessary, one methodology for each domain), then the organization should strive for a corporate-wide acceptance of that methodology for each domain. If the methodology is accepted and used only in "pockets of interest," then a fragmented culture will occur. Fragmented cultures generally do not allow the organization to maximize the benefits of project management.

OVERLAPPING LEVELS ⎯⎯⎯⎯⎯⎯⎯⎯⎯⎯⎯

Generally speaking, Levels 2 and 3 do not overlap. Once a company recognizes
the true benefits of project management and the need for a singular methodology,
the organization stops developing individual processes and focuses on what's best
for the whole.

Allowing individual processes to continue without any integration into a singular methodology gives employees a viable excuse to resist change. Employees
must be encouraged to make decisions that are in the best interest of the company
as a whole rather than in the best interest of their own department.

ASSESSMENT INSTRUMENT FOR LEVEL 3 ⎯⎯⎯⎯⎯⎯⎯

The following 42 multiple-choice questions will allow you to compare your organization against other companies with regard to the Level 3 Hexagon of
Excellence. After you complete question 42, a grading system is provided. You
can then compare your organization to some of the best who have achieved Level
3 maturity.

Please pick one and only one answer per question. A worksheet and answer
key follow the exercise.

QUESTIONS ⎯⎯⎯⎯⎯⎯⎯⎯⎯⎯⎯⎯⎯⎯⎯⎯⎯

1. My company *actively* uses the following processes:
    A. Total quality management (TQM) only
    B. Concurrent engineering (shortening deliverable development time) only
    C. TQM and concurrent engineering only
    D. Risk management only
    E. Risk management and concurrent engineering only
    F. Risk management, concurrent engineering, and TQM

2. On what percent of your projects do you use the principles of total quality management?
    A. 0 percent
    B. 5–10 percent
    C. 10–25 percent
    D. 25–50 percent
    E. 50–75 percent
    F. 75–100 percent

3. On what percent of your projects do you use the principles of risk management?
    A. 0 percent
    B. 5–10 percent
    C. 10–25 percent

    D. 25–50 percent
    E. 50–75 percent
    F. 75–100 percent

4. On what percent of your projects do you try to compress product/deliverable schedules, by performing work in parallel rather than in series?
    A. 0 percent
    B. 5–10 percent
    C. 10–25 percent
    D. 25–50 percent
    E. 50–75 percent
    F. 75–100 percent

5. My company's risk management process is based upon:
    A. We do not use risk management
    B. Financial risks only
    C. Technical risks only
    D. Scheduling risks only
    E. A combination of financial, technical, and scheduling risks based upon the project.

6. The risk management methodology in my company is:
    A. Nonexistent
    B. More informal than formal
    C. Based upon a structured methodology supported by policies and procedures
    D. Based upon a structured methodology supported by policies, procedures, and standardized forms to be completed

7. How many different project management methodologies exist in your organization (i.e., do you consider a systems development methodology for MIS projects different than a product development project management methodology)?
    A. We have no methodologies
    B. 1
    C. 2–3
    D. 4–5
    E. More than 5

8. With regard to benchmarking:
    A. My company has never tried to use benchmarking
    B. My company has performed benchmarking and implemented changes but not for project management.
    C. My company has performed project management benchmarking but no changes were made.
    D. My company has performed project management benchmarking and changes were made.

9. My company's corporate culture is best described by the concept of:
    A. Single-boss reporting
    B. Multiple-boss reporting
    C. Dedicated teams without empowerment
    D. Nondedicated teams without empowerment
    E. Dedicated teams with empowerment
    F. Nondedicated teams with empowerment

10. With regard to morals and ethics, my company believes that:
    A. The customer is always right
    B. Decisions should be made in the following sequence: best interest of the customer first, then the company, then the employees
    C. Decisions should be made in the following sequence: best interest of company first, customer second, and the employees last
    D. We have no such written policy or set of standards

11. My company conducts internal training courses on:
    A. Morality and ethics within the company
    B. Morality and ethics in dealing with customers
    C. Good business practices
    D. All of the above
    E. None of the above
    F. At least two of the first three

12. With regard to scope creep or scope changes, our culture:
    A. Discourages changes after project initiation
    B. Allows changes only up to a certain point in the project's life cycle using a formal change control process
    C. Allows changes anywhere in the project life cycle using a formal change control process
    D. Allows changes but without any formal control process

13. Our culture seems to be based upon:
    A. Policies
    B. Procedures (including forms to be filled out)
    C. Policies and procedures
    D. Guidelines
    E. Policies, procedures, and guidelines

14. Cultures are either quantitative (policies, procedures, forms, and guidelines), behavioral, or a compromise. The culture in my company is probably _____ percent behavioral.
    A. 10–25 percent
    B. 25–50 percent
    C. 50–60 percent
    D. 60–75 percent
    E. Greater than 75 percent

15. Our organizational structure is:
    A. Traditional (predominantly vertical)
    B. A strong matrix (i.e., project manager provides most of the technical direction)
    C. A weak matrix (i.e., line managers provide most of the technical direction)
    D. We use colocated teams
    E. I don't know what the structure is: management changes it on a daily basis

16. When assigned as a project leader, our project manager obtains resources by:
    A. "Fighting" for the best people available
    B. Negotiating with line managers for the best people available
    C. Negotiating for deliverables rather than people
    D. Using senior management to help get the appropriate people
    E. Taking whatever he or she gets, no questions asked

17. Our line managers:
   A. Accept total accountability for the work in their line
   B. Ask the project managers to accept total accountability
   C. Try to share accountability with the project managers
   D. Hold the assigned employees accountable
   E. We don't know the meaning of the word "accountability"; it is not part of our vocabulary.

18. In the culture within our company, the person most likely to be held accountable for the ultimate technical integrity of the final deliverable is/are:
   A. The assigned employees
   B. The project manager
   C. The line manager
   D. The project sponsor
   E. The whole team

19. In our company, the project manager's authority comes from:
   A. Within himself/herself, whatever he or she can get away with
   B. The immediate superior to the project manager
   C. Documented job descriptions
   D. Informally through the project sponsor in the form of a project charter or appointment letter

20. After project go-ahead, our project sponsors tend to:
   A. Become invisible, even when needed
   B. Micromanage
   C. Expect summary-level briefings once a week
   D. Expect summary-level briefings once every two weeks
   E. Get involved only when a critical problem occurs or at the request of the project manager or line managers.

21. What percentage of your projects have sponsors who are at the director level or above?
   A. 0–10 percent
   B. 10–25 percent
   C. 25–50 percent
   D. 50–75 percent
   E. More than 75 percent

22. My company offers approximately how many different *internal* training courses for the employees (courses that can be regarded as project-related)?
   A. Less than 5
   B. 6–10
   C. 11–20
   D. 21–30
   E. More than 30

23. With regard to the previous answer, what percentage of the courses are more behavioral than quantitative?
   A. Less than 10 percent
   B. 10–25 percent
   C. 25–50 percent
   D. 50–75 percent
   E. More than 75 percent

24. My company believes that:
   A. Project management is a part-time job
   B. Project management is a profession
   C. Project management is a profession and we should become certified as project management professionals, but at our own expense
   D. Project management is a profession and our company pays for us to become certified as project management professionals
   E. We have no project managers in our company

25. My company believes that training should be:
   A. Performed at the request of employees
   B. Performed to satisfy a short-term need
   C. Performed to satisfy both long- and short-term needs
   D. Performed only if there exists a return on investment on training dollars

26. My company believes that the content of training courses is best determined by:
   A. The instructor
   B. The Human Resource Department
   C. Management
   D. Employees who will receive the training
   E. Customization after an audit of the employees and managers

27. What percentage of the training courses in project management contain *documented* lessons learned case studies from other projects within your company?
   A. None
   B. Less than 10 percent
   C. 10–25 percent
   D. 25–50 percent
   E. More than 50 percent

28. What percentage of the executives in your functional (not corporate) organization have attended training programs or executive briefings specifically designed to show executives what they can do to help project management mature?
   A. None!
   B. Less than 25 percent
   C. 25–50 percent
   D. 50–75 percent
   E. More than 75 percent

29. In my company, employees are promoted to management because:
   A. They are technical experts
   B. They demonstrate the administrative skills of a professional manager
   C. They know how to make sound business decisions
   D. They are at the top of their pay grade
   E. Our "rank and file" pool is over its numerical upper limits

30. A report must be written and presented to the customer. Neglecting the cost to accumulate the information, the approximate cost per page for a typical report is:
   A. I have no idea
   B. $100–200 per page
   C. $200–500 per page

    D. Greater than $500 per page

    E. Free; exempt employees in our company prepare the reports at home on their own time.

31. The culture within our organization is best described as:

    A. Informal project management based upon trust, communication, and cooperation

    B. Formality based upon policies and procedures for everything

    C. Project management that thrives on formal authority relationships

    D. Executive meddling, which forces an overabundance of documentation

    E. Nobody trusting the decisions of our project managers

32. What percentage of the project manager's time each week is spent preparing reports?

    A. 5–10 percent

    B. 10–20 percent

    C. 20–40 percent

    D. 40–60 percent

    E. Greater than 60 percent

33. During project *planning,* most of our activities are accomplished using:

    A. Policies

    B. Procedures

    C. Guidelines

    D. Checklists

    E. None of the above

34. The typical time duration for a project status review meeting with senior management is:

    A. Less than 30 minutes

    B. 30–60 minutes

    C. 60–90 minutes

    D. 90 minutes–2 hours

    E. Greater than 2 hours

35. Our customers mandate that we manage our projects:

    A. Informally

    B. Formally, but with scope creep disallowed

    C. Formally, but with scope creep allowed

    D. It is our choice as long as the deliverables are met

36. My company believes that *poor* employees:

    A. Should never be assigned to teams

    B. Once assigned to a team, are the responsibility of the project manager for supervision

    C. Once assigned to a team, are the responsibility of their line manager for supervision

    D. Can be effective if assigned to the right team

    E. Should be promoted into management

37. Employees who are assigned to a project team (either full-time or part-time) have a performance evaluation conducted by:

    A. Their line manager only

    B. The project manager only

   C. Both the project and line managers
   D. Both the project and line managers, together with a review by the sponsor

38. The skills that will probably be most important for my company's project managers as we move into the twenty-first century are:
   A. Technical knowledge and leadership
   B. Risk management and knowledge of the business
   C. Integration skills and risk management
   D. Integration skills and knowledge of the business
   E. Communication skills and technical understanding

39. In my organization, the people assigned as project leaders are usually:
   A. First line managers
   B. First or second line managers
   C. Any level of management
   D. Usually nonmanagement employees
   E. Anyone in the company

40. The project managers in my organization have undergone at least some degree of training in:
   A. Feasibility studies
   B. Cost-benefit analyses
   C. Both A and B
   D. Our project managers are typically brought on board after project approval/award.

41. Our project managers are encouraged to:
   A. Take risks
   B. Take risks upon approval by senior management
   C. Take risks upon approval by project sponsors
   D. Avoid risks

42. Consider the following statement: Our project managers have a sincere interest in what happens to each team member *after* the project is scheduled to be completed.
   A. Strongly agree
   B. Agree
   C. Not sure
   D. Disagree
   E. Strongly disagree

Using the answer key that follows, please complete Exhibit 3.

ANSWER KEY _____

The assignment of the points is as follows:

Integrated Processes

Question		Points				
1	A. 2	B. 2	C. 4	D. 2	E. 4	F. 5
2	A. 0	B. 0	C. 1	D. 3	E. 4	F. 5

3	A. 0	B. 0	C. 3	D. 4	E. 5	F. 5
4	A. 0	B. 1	C. 3	D. 4	E. 5	F. 5
5	A. 0	B. 2	C. 2	D. 2	E. 5	
6	A. 0	B. 2	C. 4	D. 5		
7	A. 0	B. 5	C. 4	D. 2	E. 0	

Culture

Question			Points			
8	A. 0	B. 2	C. 3	D. 5		
9	A. 1	B. 3	C. 4	D. 4	E. 5	F. 5
10	A. 1	B. 5	C. 4	D. 0		
11	A. 3	B. 3	C. 3	D. 5	E. 0	F. 4
12	A. 1	B. 5	C. 5	D. 3		
13	A. 2	B. 3	C. 4	D. 5	E. 4	
14	A. 2	B. 3	C. 4	D. 5	E. 5	

Management Support

Question			Points		
15	A. 1	B. 5	C. 5	D. 5	E. 0
16	A. 2	B. 3	C. 5	D. 0	E. 2
17	A. 4	B. 2	C. 5	D. 1	E. 0
18	A. 2	B. 3	C. 5	D. 0	E. 3
19	A. 4	B. 1	C. 2	D. 5	
20	A. 1	B. 1	C. 3	D. 4	E. 5
21	A. 1	B. 3	C. 5	D. 4	E. 4

Training and Education

Question			Points		
22	A. 1	B. 3	C. 5	D. 5	E. 5
23	A. 0	B. 2	C. 4	D. 5	E. 5
24	A. 0	B. 3	C. 4	D. 5	E. 0
25	A. 2	B. 3	C. 4	D. 5	
26	A. 2	B. 1	C. 2	D. 3	E. 5
27	A. 0	B. 1	C. 3	D. 5	E. 5
28	A. 0	B. 1	C. 3	D. 4	E. 5

Informal Project Management

Question			Points		
29	A. 2	B. 4	C. 5	D. 1	E. 0
30	A. 0	B. 3	C. 4	D. 5	E. 0
31	A. 5	B. 2	C. 3	D. 1	E. 0
32	A. 3	B. 5	C. 4	D. 2	E. 1
33	A. 2	B. 3	C. 4	D. 5	E. 0
34	A. 4	B. 5	C. 3	D. 1	E. 0
35	A. 3	B. 4	C. 3	D. 5	

Behavioral Excellence

Question			Points		
36	A. 1	B. 2	C. 4	D. 5	E. 0
37	A. 5	B. 1	C. 4	D. 2	
38	A. 3	B. 5	C. 5	D. 5	E. 4
39	A. 2	B. 2	C. 2	D. 5	E. 3
40	A. 3	B. 3	C. 5	D. 1	
41	A. 5	B. 3	C. 4	D. 1	
42	A. 5	B. 4	C. 2	D. 1	E. 1

Exhibit 3

Determine your points for each of the questions and complete the following:

A. Points for integrated processes (Questions 1–7):   _____

B. Points for culture (Questions 8–14):   _____

C. Points for management support (Questions 15–21):   _____

D. Points for training and education (Questions 22–28):   _____

E. Points for informal project management (Questions 29–35): _____

F. Points for behavioral excellence (Questions 36–42):   _____

                                          TOTAL: _____

EXPLANATION OF POINTS FOR LEVEL 3

Each of the six areas are components of the Hexagon of Excellence discussed in Level 3. The total points can be interpreted as follows:

Points	*Interpretation*
169–210	Your company compares very well to the companies discussed in this text. You are on the right track for excellence, assuming that you have not achieved it yet. Continuous improvement will occur.
147–168	Your company is going in the right direction, but more work is still needed. Project management is not totally perceived as a profession. It is also possible that your organization simply does not fully understand project management. Emphasis is probably more toward being non–project-driven than project-driven.
80–146	The company is probably just providing lip service to project manage-ment. Support is minimal. The company believes that it is the right thing to do, but has not figured out the true benefits or what they, the execu-tives, should be doing. The company is still a functional organization.
Below 80	Perhaps you should change jobs or seek another profession. The com-pany has no understanding of project management, nor does it appear that the company wishes to change. Line managers want to maintain their existing power base and may feel threatened by project management.

Level 4:
Benchmarking

INTRODUCTION

Project management benchmarking is the process of continuously comparing the project management practices of your organization with the practices of leaders anywhere in the world; its goal is to to gain information to help you improve your own performance. The information obtained through benchmarking might be used to help you improve your processes and the way in which those processes are executed, or the information might be used to help your company become more competitive in the marketplace.

Benchmarking is a continuous effort of analysis and evaluation. Care must be taken in deciding what to benchmark. It is impossible and impractical to evaluate every aspect of project management. It is best to decide on those few critical success factors that must go right for your business to flourish. For project management benchmarking, the critical success factors are usually the key business processes and how they are integrated. If these key success factors do not exist, then the organization's efforts may be hindered.

Deciding what information to benchmark against is usually easier than obtaining that information. Locating some information will require a critical search. Some information may be hard to find. Some information you would find helpful might not be available for release because the organization that has it views it as proprietary. Identifying the target companies against which you should benchmark may not be as easy as you believe.

Benchmarking has become common since it was first popularized by Xerox during the 1980s. Benchmarking is an essential ingredient for those companies that have won the prestigious Malcolm Baldrige Award. Most of these award winners seem willing to readily share their project management experiences. Unfortunately, there are some truly excellent companies in project management that have not competed for these awards because they do not want their excellence displayed.

Benchmarking for project management can be accomplished through surveys, questionnaires, attending local chapter meetings of the Project Management Institute (PMI), and attending conferences and symposiums. Personal contacts often provide the most valued sources of information.

There is a so-called "Code of Conduct" for benchmarking:

- Keep the benchmarking process legal.
- Do not violate rules of confidentiality.
- Sharing information is a two-way street.
- Be willing to sign a nondisclosure form.
- Do not share *any* information received with a third party without written permission.
- Emphasize guidelines and checklists but avoid asking for forms that may be highly sensitive.

Benchmarking should not be performed unless your organization is willing to make changes. The changes must be part of a structured process that includes evaluation, applicability, and risk management. Benchmarking is part of the strategic planning process for project management that results in an action plan ready for implementation.

CHARACTERISTICS

Level 4 is the level where the organization realizes that its existing methodology can be improved upon. The complexity rests in figuring out how to achieve that improvement. For project-driven companies, continuous improvement is a means to maintain or improve upon a competitive advantage. Continuous improvement is best accomplished through continuous benchmarking. The company must decide whom to benchmark and what to benchmark.

There are certain characteristics of Level 4, as show in Figure 8–1:

- The organization must establish a project office (PO) or a center of excellence (COE) for project management. This is the focal position in the company for project management knowledge.
- The PO or COE must be dedicated to the project management improvement process. This is usually accomplished with full-time, dedicated personnel.
- Benchmarking must be made against both similar and nonsimilar industries. In today's world, a company with five years of experience in proj-

Benchmarking

- Establishment of a Project Office (PO) or a Center of Excellence (COE)
- Dedication to Benchmarking
- Looking at Both Similar and Nonsimilar Industries
- Quantitative Benchmarking (Processes and Methodologies)
- Qualitative Benchmarking (Cultures)

FIGURE 8–1. Characteristics of Level 4.

ect management could easily surpass the capabilities of a company that has used project management for 20 years or more.

- The company should perform both quantitative and qualitative benchmarking. Quantitative benchmarking analyzes processes and methodologies, whereas qualitative benchmarking looks at project management applications.

THE PROJECT OFFICE/ CENTER OF EXCELLENCE

When companies reach Level 4, they are committed to project management across the entire organization. Project management knowledge is now considered as essential for the survival of the firm. To centralize the knowledge on project management, organizations have created a project office (PO) or a center of excellence (COE) for project management.

Responsibilities for a PO/COE include:

- A strategic planning focal point of project management
- An organization dedicated to benchmarking for project management
- An organization dedicated to continuous improvement
- An organization that provides mentorship for inexperienced project managers
- A centralized data bank on lessons learned

TABLE 8–1. PROJECT OFFICE VERSUS CENTER OF EXCELLENCE

Project Office	Center of Excellence
• Permanent line function for project manager • Focus on internal lessons learned activities • Champion for the implementation of the methodology • Expertise in the use of project management tools	• May be a formal or informal committee (may be part-time) • Focuses on external bench-marking • Champion for continuous improvement and benchmarking • Expertise in the identification of project management tools

- An organization for sharing project management ideas and experiences
- A "hot line" for problem-solving that does not automatically inform senior management
- An organization for creating project management standards
- A focal point for centralized planning and scheduling activities
- A focal point for centralized cost control and reporting
- An organization to assist Human Resources in the creation of a project management career path
- An organization to assist Human Resources in developing a project management curriculum

Most companies view the PO and the COE as being two names for the same thing. There are, however, fundamental differences, as shown in Table 8–1. Despite the responsibilities, companies are struggling with the organizational reporting location of the PO/COE. There appears to be agreement that the location should be at the senior levels of management. Figure 8–2 shows a simplified organizational chart for a PO.

FIGURE 8–2. Simplified PO organizational chart.

BENCHMARKING OPPORTUNITIES

Historically, benchmarking is accomplished by two approaches: competitive benchmarking and process benchmarking. Competitive benchmarking concentrates on deliverables and quantitative critical success factors. Process benchmarking focuses on process performance and functionality. Process benchmarking is most closely aligned to project management. For simplicity's sake, we will consider only process improvement benchmarking. We can break it down into quantitative (i.e., integration) process improvement opportunities and qualitative process improvement opportunities.

Figure 8–3 shows the quantitative process improvement opportunities, which center around enhancements due to integration opportunities. The five major areas identified in Figure 8–3 are the five integrated processes described in Level 3 of the project management maturity model (PMMM).

Figure 8–4 shows the qualitative process improvement opportunities, which center around applications and further changes to the corporate culture. Included in the qualitative process improvement activities are:

- Corporate acceptance: This includes getting the entire organization to accept a singular methodology for managing projects. Pockets of project management support tend to hinder rapid acceptance of project management by the rest of the organization. To obtain corporate acceptance, we must:
  - Increase the usage and support of existing users

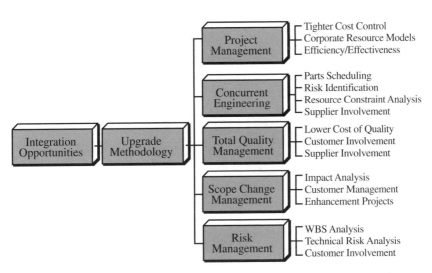

FIGURE 8–3. Quantitative process improvement opportunities (generic integrated process strategies).

- Attract new internal users, those who have been providing resistance to project management
- Discourage the development of parallel methodologies, which can create further pockets of project management. This is done by showing the added costs of parallelization.
- Emphasize the present and future benefits to the corporation that will result from using a singular methodology.

● Integrated processes: This is a recognition that the singular methodology can be enhanced further by integrating other existing processes into the singular methodology. Typically, this includes business processes such as capital budgeting, feasibility studies, cost-benefit analyses, and return-on-investment analyses. New processes that could be integrated include supply chain management.

● Enhanced benchmarking: Everyone tends to benchmark against the best within their own industry, but benchmarking against nonsimilar industries can be just as fruitful. An aerospace company spent over ten years benchmarking only against other aerospace companies. During the mid-1990s, the firm began benchmarking against nonaerospace firms, and found that these firms had developed outstanding methodologies with capabilities exceeding those of the aerospace firm.

● Software enhancements: Although off-the-shelf software packages exist, most firms still need some type of customization. This can be done through internal upgrades for customization or by new purchases, with the software vendor developing the customization.

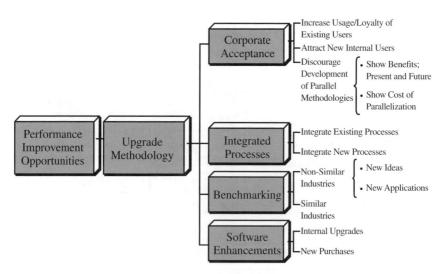

FIGURE 8–4. Qualitative process improvement opportunities (generic performance improvement strategies).

ROADBLOCKS

There also exist roadblocks to completing Level 4 and reaching Level 5, as seen in Figure 8–5. The singular methodology created in Level 3 was developed internally within the company. Benchmarking may indicate that improvements can be made. The original architects of the singular methodology may resist change with arguments such as: "It wasn't invented here," or "It does not apply to us." Another form of resistance is the argument that we have benchmarked against the wrong industry.

People are inherently fearful of change, and benchmarking opens the door for unexpected results to surface. Sooner or later everyone realizes that benchmarking is a necessity for company survival. It is at this junction that a serious commitment to benchmarking occurs.

ADVANCEMENT CRITERIA

There are four key actions required by the organization to advance to Level 5, the final level. These actions are as follows:

- Create an organization dedicated to benchmarking.
- Develop a project management benchmarking process.
- Decide what to benchmark and against whom to benchmark.
- Recognize the benefits of benchmarking.

FIGURE 8–5. Roadblocks to completion of Level 4.

The successful completion of Level 4 is accompanied by a low degree of difficulty. Since the organization has already accepted the idea of a singular methodology, it is a low risk to expect the employees to accept changes. They now know that change is inevitable.

ASSESSMENT INSTRUMENT FOR LEVEL 4

On the next several pages you will find 25 questions concerning how mature you believe your organization to be. Beside each question you will circle the number that corresponds to your opinion. In the example below, your choice would have been "Slightly Agree."

−3	Strongly Disagree
−2	Disagree
−1	Slightly Disagree
0	No Opinion
(+1)	Slightly Agree
+2	Agree
+3	Strongly Agree

Example: (−3, −2, −1, 0, (+1), +2, +3)

The row of numbers from −3 to +3 will be used later for evaluating the results. After answering Question 25, you will grade the exercise.

QUESTIONS

The following 25 questions involve *benchmarking*. Please answer each question as honestly as possible. Circle the answer you feel is correct, *not* the answer you believe the question is seeking out.

1. Our benchmarking studies have found companies with tighter cost control processes. (−3 −2 −1 0 +1 +2 +3)

2. Our benchmarking studies have found companies with better *impact analysis* during scope change control. (−3 −2 −1 0 +1 +2 +3)

3. Our benchmarking studies have found that companies are performing risk management by analyzing the detailed level of the work breakdown structure (WBS). (−3 −2 −1 0 +1 +2 +3)

4. Our benchmarking studies are investigating *supplier involvement* in project management activities. $(-3 \quad -2 \quad -1 \quad 0 \quad +1 \quad +2 \quad +3)$

5. Our benchmarking studies are investigating *customer involvement* in project management activities. $(-3 \quad -2 \quad -1 \quad 0 \quad +1 \quad +2 \quad +3)$

6. Our benchmarking studies are investigating how to obtain increased loyalty/usage of our project management methodology. $(-3 \quad -2 \quad -1 \quad 0 \quad +1 \quad +2 \quad +3)$

7. Our benchmarking efforts are looking at industries in the same business area as our company. $(-3 \quad -2 \quad -1 \quad 0 \quad +1 \quad +2 \quad +3)$

8. Our benchmarking efforts are looking at nonsimilar industries (i.e., industries in different business areas). $(-3 \quad -2 \quad -1 \quad 0 \quad +1 \quad +2 \quad +3)$

9. Our benchmark efforts are looking at nonsimilar industries to seek out new ideas and new applications for project management. $(-3 \quad -2 \quad -1 \quad 0 \quad +1 \quad +2 \quad +3)$

10. Our benchmarking efforts are looking at other company's concurrent engineering activities to see how they perform *parts* scheduling and tracking. $(-3 \quad -2 \quad -1 \quad 0 \quad +1 \quad +2 \quad +3)$

11. Our benchmarking efforts have found other companies that are performing *resource constraint* analyses. $(-3 \quad -2 \quad -1 \quad 0 \quad +1 \quad +2 \quad +3)$

12. Our benchmarking efforts are looking at the way other companies manage their *customers* during the scope change management process. $(-3 \quad -2 \quad -1 \quad 0 \quad +1 \quad +2 \quad +3)$

13. Our benchmarking efforts are looking at the way other companies involve their *customers* during risk management activities. $(-3 \quad -2 \quad -1 \quad 0 \quad +1 \quad +2 \quad +3)$

14. Our benchmarking efforts are looking at software enhancements through *internal upgrades*. $(-3 \quad -2 \quad -1 \quad 0 \quad +1 \quad +2 \quad +3)$

15. Our benchmarking efforts are looking at software enhancements through *new purchases*. $(-3 \quad -2 \quad -1 \quad 0 \quad +1 \quad +2 \quad +3)$

16. Our benchmarking efforts are looking at the way other companies attract new, internal users to their methodology for project management. $(-3 \quad -2 \quad -1 \quad 0 \quad +1 \quad +2 \quad +3)$

17. Our benchmarking efforts are focusing
    on how other companies perform
    technical risk management. $(-3 \quad -2 \quad -1 \quad 0 \quad +1 \quad +2 \quad +3)$

18. Our benchmarking efforts are focusing
    on how other companies obtain better
    efficiency and effectiveness of their
    project management methodology. $(-3 \quad -2 \quad -1 \quad 0 \quad +1 \quad +2 \quad +3)$

19. Our benchmarking efforts focus on how
    to obtain a lower cost of quality. $(-3 \quad -2 \quad -1 \quad 0 \quad +1 \quad +2 \quad +3)$

20. Our benchmarking efforts are looking at
    the way other companies are performing
    risk management during *concurrent
    engineering* activities. $(-3 \quad -2 \quad -1 \quad 0 \quad +1 \quad +2 \quad +3)$

21. Our benchmarking efforts are looking at
    the way other companies use
    enhancement projects as part of scope
    change management. $(-3 \quad -2 \quad -1 \quad 0 \quad +1 \quad +2 \quad +3)$

22. Our benchmarking efforts are looking at
    ways of integrating *existing processes*
    into our singular methodology. $(-3 \quad -2 \quad -1 \quad 0 \quad +1 \quad +2 \quad +3)$

23. Our benchmarking efforts are looking at
    ways other companies have integrated
    new methodologies and processes into
    their singular methodology. $(-3 \quad -2 \quad -1 \quad 0 \quad +1 \quad +2 \quad +3)$

24. Our benchmarking efforts are looking at
    the way other companies handle or
    discourage the development of *parallel*
    methodologies. $(-3 \quad -2 \quad -1 \quad 0 \quad +1 \quad +2 \quad +3)$

25. Our benchmarking efforts are seeking
    out other companies' use of *corporate
    resource models.* $(-3 \quad -2 \quad -1 \quad 0 \quad +1 \quad +2 \quad +3)$

An answer sheet to complete follows. Please complete Exhibit 4.

Exhibit 4

Each response you circled in Questions 1–25 had a column value between -3 and $+3$. In the appropriate spaces below, place the circled value (between -3 and $+3$) beside each question.

Quantitative Benchmarking

1. _____
2. _____
3. _____
4. _____
5. _____
10. _____
11. _____
12. _____
13. _____
17. _____
18. _____
19. _____
20. _____
21. _____
25. _____

TOTAL _____

Qualitative Benchmarking

6. _____
7. _____
8. _____
9. _____
14. _____
15. _____
16. _____
22. _____
23. _____
24. _____

TOTAL _____

Quantitative benchmarking total: _____

Qualitative benchmarking total: _____

Combined total: _____

EXPLANATION OF POINTS FOR LEVEL 4 _____

This exercise measures two items: Is your organization performing benchmarking and, if so, are you emphasizing quantitative or qualitative benchmarking?

Quantitative benchmarking investigates improvements to the methodology and processes. Scores greater than 25 are excellent and imply that your organization is committed to quantitative benchmarking. Scores less than 10 indicate a lack of commitment or that the organization does not understand how to benchmark or against whom to benchmark. Scores between 11 and 24 indicate that some benchmarking may be taking place, but a PO or COE is not in place as yet.

Qualitative benchmarking looks more at applications benchmarking and how the culture executes the methodology. Scores greater than 12 are excellent. Scores less than 5 indicate that not enough emphasis is placed upon the "soft side" of benchmarking. Scores between 6 and 11 are marginally acceptable.

Combined scores (i.e., quantitative and qualitative) of 37 or more imply that your organization is performing benchmarking well. The right information is being considered and the right companies are being targeted. The balance between quantitative and qualitative benchmarking is good. The company probably has a COE or PO in place.

Level 5: Continuous Improvement

CHARACTERISTICS

In the previous level, the organization began benchmarking against other companies. In Level 5, the organization evaluates the information learned during benchmarking and *implements* the changes necessary to improve the project management process. It is in this level that the company comes to the realization that excellence in project management is a never-ending journey.

There are four characteristics of Level 5, as shown in Figure 9–1.

- The organization must create lessons learned files from the debriefing sessions at the end of each project. Case studies on each project, discussing mistakes made and knowledge learned, are critical so that mistakes are not repeated.
- The knowledge learned on each project must be transferred to other projects and teams. This can be accomplished through quarterly or semiannual lessons learned forums or from lessons learned case studies discussed in training programs.
- The company must recognize that a mentorship program should be put in place to groom future project managers. Knowledge transfer and lessons learned information can be transmitted through the mentorship program as well. The mentorship program is best administered through a Project Office (PO) or a Center of Excellence (COE).
- The final characteristic of Level 5 is a corporate-wide understanding that strategic planning for project management is a continuous, ongoing process.

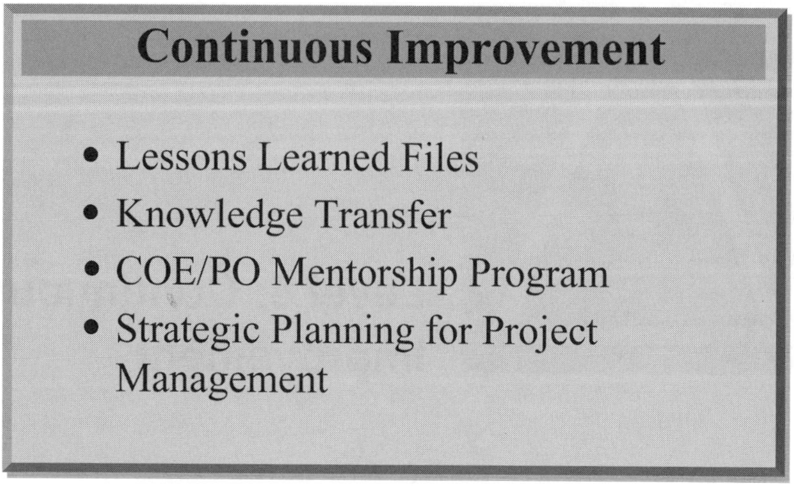

> # Continuous Improvement
>
> - Lessons Learned Files
> - Knowledge Transfer
> - COE/PO Mentorship Program
> - Strategic Planning for Project Management

FIGURE 9–1. Characteristics of Level 5.

Documenting project results in lessons learned files and the preparation of case studies can be difficult to implement. People learn from both successes and failures. One executive commented that the only true project failures are the ones from which we learned nothing. Another executive commented that project de-briefings are a waste of time unless we learn something from them.

Documenting successes is easy. Documenting mistakes is more troublesome because people do not want their names attached to mistakes for fear of retribution. Company employees still know which individuals worked on which projects, even when the case study is disguised. A strong corporate culture is needed to make documenting mistakes work effectively.

CONTINUOUS IMPROVEMENT AREAS

Project management methodologies must undergo continuous improvement. This may be strategically important to stay ahead of the competition. Continuous improvements to a methodology can be internally driven by factors such as better software availability, a more cooperative corporate culture, or simply training and education in the use of the methodology. Externally driven factors include relationships with customers and suppliers, legal factors, social factors, technological factors, and even political factors.

Five areas for continuous improvement to the project management methodology are shown in Figure 9–2 and in the following:

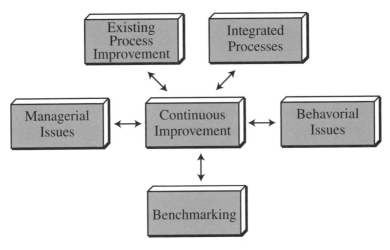

FIGURE 9–2. Factors to consider for continuous improvement.

Existing Process Improvements
- Frequency of use: Has prolonged use of the methodology made it apparent that changes can be made?
- Access to customers: Can we improve the methodology to get closer to our customers?
- Substitute products: Are there new products (i.e., software) in the marketplace that can replace and improve part of our methodology?
- Better working conditions: Can changes in the working conditions cause us to eliminate parts of the methodology (i.e., paperwork requirements)?
- Better use of software: Will new or better use of the software allow us to eliminate some of our documentation and reports?

Integrated Process Improvements
- Speed of integration: Are there ways to change the methodology to increase the speed of integrating activities?
- Training requirements: Have changes in our training requirements mandated changes in our methodology?
- Corporate-wide acceptance: Should the methodology change in order to obtain corporate-wide acceptance?

Behavioral Issues
- Changes in organizational behavior: Have changes in behavior mandated methodology changes?
- Cultural changes: Has our culture changed (i.e., to a cooperative culture) such that the methodology can be enhanced?

- Management support: Has management support improved to a point where fewer gate reviews are required?
- Impact on informal project management: Is there enough of a cooperative culture such that informal project management can be used to execute the methodology?
- Shifts in power and authority: Do authority and power changes mandate a looser or a more rigid methodology?
- Safety considerations: Have safety or environmental changes occurred that will impact the methodology?
- Overtime requirements: Do new overtime requirements mandate an updating of forms, policies, or procedures?

Benchmarking

- Creation of a project management COE: Do we now have a "core" group responsible for benchmarking?
- Cultural benchmarking: Do other organizations have better cultures than we do in project management execution?
- Process benchmarking: What new processes are other companies integrating into their methodology?

Managerial Issues

- Customer communications: Have there been changes in the way we communicate with our customers?
- Resource capability versus needs: If our needs have changed, what has happened to the capabilities of our resources?
- Restructuring requirements: Has restructuring caused us to change our sign-off requirements?
- Growing pains: Does the methodology have to be updated to include our present growth in business (i.e., tighter or looser controls)?

The five factors considered above provide a company with a good framework for continuous improvement. The benefits of continuous improvement include:

- Better competitive positioning
- Corporate unity
- Improved cost analysis
- Customer value added
- Better management of customer expectations
- Ease of implementation

THE NEVER-ENDING CYCLE

Given the fact that maturity in project management is a never-ending journey, we can define excellence in project management as a never-ending cycle of bench-

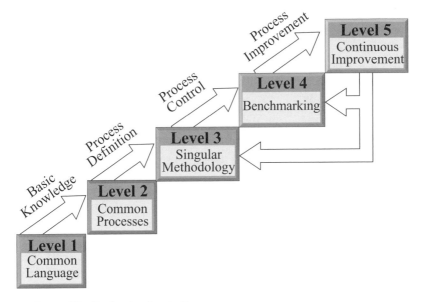

FIGURE 9–3. The five levels of maturity.

marking–continuous improvement–singular methodology enhancement, as shown in Figure 9–3. This implies that Levels 3, 4, and 5 of the PMMM are repeated over and over again. This also justifies our statement of the need for overlapping levels.

EXAMPLES OF CONTINUOUS IMPROVEMENT

As more and more industries accept project management as a way of life, the continuous improvement opportunities in project management practices have arisen at an astounding rate. What is even more important is the fact that companies are sharing their accomplishments with other companies during benchmarking activities.

Ten recent interest areas are included in this chapter:

- Developing effective procedural documentation
- Project management methodologies
- Continuous improvement
- Capacity planning
- Competency models
- Managing multiple projects
- End-of-phase review meetings
- Strategic selection of projects

- Portfolio selection of projects
- Horizontal accounting

These ten topics appear to be the quickest to change as we enter the twenty-first century.

DEVELOPING EFFECTIVE PROCEDURAL DOCUMENTATION

Previously, we showed the necessity to develop processes and ultimately a singular methodology for project management. Project management methodologies require a project management information system (PMIS), which is based upon procedural documentation. The procedural documentation can be in the form of policies, procedures, guidelines, forms and checklists, or even a combination of these. Good procedural documentation will accelerate the project management maturity process, foster support at all levels of management, and greatly improve project communications. The type of procedural documentation selected can change over the years and is heavily biased on whether we wish to manage more formally or informally. In any event, procedural documentation supports effective communications, which in turn, provides for better interpersonal skills.

An important facet of any project management methodology is to provide the people in the organization with procedural documentation on how to conduct project-oriented activities and how to communicate in such a multidimensional environment. The project management policies, procedures, forms, and guidelines can provide some of these tools for delineating the process, as well as a format for collecting, processing, and communicating project-related data in an orderly, standardized format. Project planning and tracking, however, involve more than just the generation of paperwork. They require the participation of the entire project team, including support departments, subcontractors, and top management. This involvement of the entire team fosters a unifying team environment. This unity, in turn, helps the team focus on the project goals and, ultimately, fosters each team member's personal commitment to accomplishing the various tasks within time and budget constraints. The specific benefits of procedural documents, including forms and checklists, are that they help to:

- Provide guidelines and uniformity
- Encourage useful, but minimum, documentation
- Communicate clearly and effectively
- Standardize data formats
- Unify project teams
- Provide a basis for analysis
- Document agreements for future reference
- Refuel commitments

- Minimize paperwork
- Minimize conflict and confusion
- Delineate work packages
- Bring new team members onboard
- Build an experience track and method for future projects

Done properly, the process of project planning must involve both the performing and the customer organizations. This involvement creates a new insight into the intricacies of a project and its management methods. It also leads to visibility of the project at various organizational levels, management involvement, and support. It is this involvement at all organizational levels that stimulates interest in the project and the desire for success, and fosters a pervasive reach for excellence that unifies the project team. It leads to commitment toward establishing and reaching the desired project objectives and to a self-forcing management system where people want to work toward these established objectives.

The Challenges

Despite all these benefits, management is often reluctant to implement or fully support a formal project management system. Management concerns often center around four issues: overhead burden, start-up delays, stifled creativity, and reduced self-forcing control. First, the introduction of more organizational formality via policies, procedures, and forms might cost some money, plus additional funding will be needed to support and maintain the system. Second, the system is seen, especially by action-oriented managers, as causing undesirable start-up delays by requiring the putting of certain stakes into the ground, in terms of project definition, feasibility, and organization, before the detailed implementation can start. Third and fourth, the system is often perceived as stifling creativity and shifting project control from the responsible individual to an impersonal process that enforces the execution of a predefined number of procedural steps and forms without paying attention to the complexities and dynamics of the individual project and its possibly changing objectives.

The comment of one project manager may be typical for many situations: "My support personnel feels that we spend too much time planning a project up front; it creates a very rigid environment that stifles innovation. The only purpose seems to be establishing a basis for controls against outdated measures and for punishment rather than help in case of a contingency." This comment is echoed by many project managers. It's not a groundless attitude either, for it also illustrates a potential misuse of formal project management systems, establishment of unrealistic controls and penalties for deviations from the program plan rather than help in finding solutions. Whether these fears are real or imaginary within a particular organization does not change the situation. It is the perceived coercion that leads to the rejection of the project management system. An additional concern is the lack of management involvement and funding to implement the project management system. Often the customer or sponsor organization must also be involved and agree with the process for planning and controlling the project.

How to Make It Work

Few companies have introduced project management procedures with ease. Most have experienced problems ranging from skepticism to sabotage of the procedural system. Realistically, however, program managers do not have much of a choice, especially for larger, more complex programs. Every project manager who believes in project management has his or her own success story. It is interesting to note, however, that many have had to use incremental approaches to develop and implement their project management methodology.

Developing and implementing such a methodology incrementally is a multi-faceted challenge to management. The problem is seldom one of understanding the techniques involved, such as budgeting and scheduling, but rather one of involving the project team in the process, getting their input, support, and commitment, and establishing a supportive environment. Furthermore, project personnel must have the feeling that the policies and procedures of the project management system facilitate communication, are flexible and adaptive to the changing environment, and provide an early warning system through which project personnel can obtain assistance rather than punishment in case of a contingency.

The procedural guidelines and forms of an established project management methodology can be especially useful during the project planning/definition phase. Not only do they help to delineate and communicate the four major sets of variables for organizing and managing the project—(1) tasks, (2) timing, (3) resources, and (4) responsibilities—they also help to define measurable milestones, as well as report and review requirements. This in turn makes it possible to measure project status and performance and supplies the crucial inputs for controlling the project toward the desired results.

Developing an effective project management methodology takes more than just a set of policies and procedures. It requires the integration of these guidelines and standards into the culture and value system of the organization. Management must lead the overall efforts and foster an environment conducive to teamwork. The greater the team spirit, trust, commitment and quality of information exchange among team members, the more likely it is that the team will develop effective decision-making processes, make individual and group commitments, focus on problem-solving, and operate in a self-forcing, self-correcting control mode. These are the characteristics that will support and pervade the formal project management system and make it work for you. When understood and accepted by the team members, such a system provides the formal standards, guidelines, and measures needed to direct a project toward specific results within the given time and resource constraints.

Established Practices

Although project managers may have the right to establish their own policies and procedures, many companies have taken the route of designing project control forms that can be used uniformly on all projects to assist in the communications process. Project control forms serve two vital purposes by establishing a common framework from which:

- The project manager will communicate with executives, functional managers, functional employees, and clients
- Executives and the project manager can make meaningful decisions concerning the allocation of resources.

Success or failure of a project depends upon the ability of key personnel to have sufficient data for decision-making. Project management is often considered to be both an art and a science. It is an art because of the strong need for interpersonal skills, and the project planning and control forms attempt to convert part of the "art" into a science.

Many companies tend not to realize until too late the necessity of good planning and control forms. Today, some of the larger companies with mature project management structures maintain a separate functional unit for forms control. This is quite common in aerospace and defense, but is also becoming common practice in other industries. Yet some executives still believe that forms are needed only when the company grows to a point where a continuous stream of unique projects necessitates some sort of uniform control mechanism.

In some small or non–project-driven organizations, each project can have its own forms. But for most other organizations, uniformity is a must. Quite often, the actual design and selection of the forms is made by individuals other than the users. This can easily lead to disaster.

Large companies with a multitude of different projects do not have the luxury of controlling projects with three or four forms. There are different forms for planning, scheduling, controlling, authorizing work, and so on. It is not uncommon for companies to have 20 to 30 different forms, each dependent upon the type of project, length of project, dollar value, type of customer reporting, and other such factors.

In project management, the project manager is often afforded the luxury of being able to set up his or her own administration for the project, a fact that could lead to irrevocable long-term damage if each project manager were permitted to design his or her own forms for project control. Many times this problem remains unchecked, and the number of forms grows exponentially with each project.

Executives can overcome this problem either by limiting the number of forms necessary for planning, scheduling, and controlling projects, or by establishing a separate department to develop the needed forms. Neither of these approaches is really practical or cost-effective. The best method appears to be the task force concept, where both managers and doers will have the opportunity to interact and provide input. In the short run, this may appear to be ineffective and a waste of time and money. However, in the long run there should be large benefits.

To be effective, the following ground rules can be used:

- Task forces should include managers as well as doers.
- Task force members must be willing to accept criticism from other peers, superiors, and especially subordinates who must "live" with these forms.

- Upper level management should maintain a rather passive (or monitoring) involvement.
- A minimum of signature approvals should be required for each form.
- Forms should be designed so that they can be updated periodically.
- Functional managers and project managers must be dedicated and committed to the use of the forms.

Categorizing the Broad Spectrum of Documents

The dynamic nature of project management and its multifunctional involvement create a need for a multitude of procedural documents to guide a project through the various phases and stages of integration. Especially for larger organizations, the challenge is not only to provide management guidelines for each project activity, but also to provide a coherent procedural framework within which project leaders from all disciplines can work and communicate with each other. Specifically, each policy or procedure must be consistent with and accommodating to the various other functions that interface with the project over its life cycle. This complexity of intricate relations is illustrated in Figure 9–4.

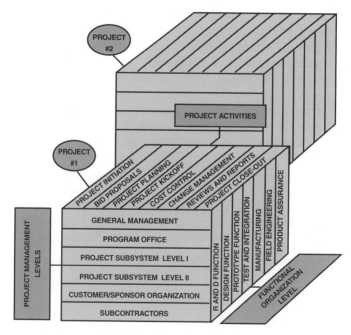

FIGURE 9–4. Interrelationship of project activities with various functional/organizational levels and project management levels. *Source:* Reprinted from H. Kerzner and H. J. Thamhain, *Project Management Operating Guidelines.* New York: Van Nostrand Reinhold, 1985.

One simple and effective way of categorizing the broad spectrum of procedural documents is by utilizing the work breakdown concept, as shown in Figure 9–5. This concept organizes the principal procedural categories along the lines of the principal project life cycle phases. Each category is then subdivided into (1) general management guidelines, (2) policies, (3) procedures, (4) forms, and (5) checklists. If necessary, the concept can be extended an additional step to develop policies, procedures, forms, and checklists for the various project and functional sublevels of operation. Although this level of formality might be needed for very large programs, an effort should be made to minimize "layering" of policies and procedures as the additional bureaucracy can cause new interface problems and additional overhead costs. For most projects, a single document covers all levels of project operations.

As We Mature . . .

As companies become more mature in executing the project management methodology, project management policies and procedures are discarded and replaced with guidelines, forms, and checklists. More flexibility is thus provided the project manager. Unfortunately, reaching this stage takes time, because executives need to develop confidence in the ability of the project management methodology to work without the rigid controls provided by policies and procedures. All companies seem to go through the evolutionary stage of relying on policies and procedures before they advance to guidelines, forms, and checklists.

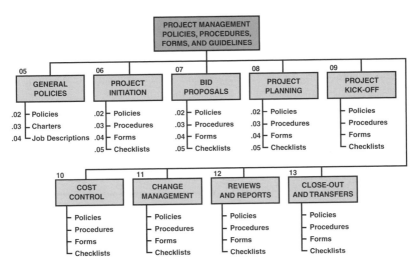

FIGURE 9–5. Categorizing procedural documents within a work breakdown structure.
Source: Reprinted from H. Kerzner and H. J. Thamhain, *Project Management Operating Guidelines.* New York: Van Nostrand Reinhold, 1985.

PROJECT MANAGEMENT
METHODOLOGIES

The ultimate purpose of any project management system is to drastically increase the likelihood that your organization will have a continuous stream of success-fully managed projects. The best way to achieve this goal is with the development of a project management methodology. Good project management methodologies are based upon guidelines and forms rather than policies and procedures. Methodologies must have enough flexibility such that they can be adapted easily to each and every project. There are consulting companies out there who have created their own methodologies and who will try to convince you that the solu-tion to most of your project management problems can be resolved with the pur-chase of their (often expensive) methodology. The primary goal of these consult-ing companies is turning problems into gold: *your problems into their gold!*

One major hurdle that any company must overcome when developing or pur-chasing a project management methodology is the fact that a methodology is nothing more than a sheet of paper with instructions. To convert this sheet of pa-per into a successful methodology, the company must accept, support, and exe-cute the methodology. If this is going to happen, the methodology should be de-signed to support the corporate culture, not vice versa. It is a fatal mistake to purchase a canned methodology package that mandates that you change your cor-porate culture to support it. If the methodology does not support the culture, the result will be a lack of acceptance of the methodology, sporadic use at best, in-consistent application of the methodology, poor morale, and perhaps even dimin-ishing support for project management. What converts any methodology into a world-class methodology is its adaptability to the corporate culture.

There is no reason why organizations cannot develop their own methodolo-gies. Companies such as Johnson Controls and Motorola are regarded as having world-class methodologies for project management and, in each case, the meth-odology was developed internally. The amount of time and effort needed to develop a methodology will vary from company to company, based upon such factors as the size and nature of the projects, the number of functional boundaries to be crossed, whether the organization is project-driven or non–project-driven, and competitive pressures.

CONTINUOUS IMPROVEMENT

All too often complacency directs the decision-making process. This is particu-larly true of organizations that have reached some degree of excellence in project management and become self-satisfied. They often realize only too late that they have lost their competitive advantage. This occurs when organizations fail to rec-ognize the importance of continuous improvement.

Figure 9–6 illustrates the need for continuous improvement. As companies

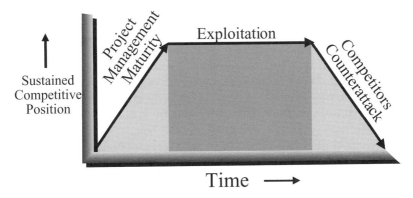

FIGURE 9–6. The need for continuous improvement.

begin to mature in project management and reach some degree of excellence, they achieve a sustained competitive advantage. Achieving this edge might very well be the single most important strategic objective of the firm. Once the firm has this sustained competitive advantage, it will then begin to exploit it.

Unfortunately, the competition will not be sitting by idly watching you exploit your sustained competitive advantage; they will begin to counterattack. When they do, you may lose a large portion, if not all, of your sustained competitive advantage. To remain effective and competitive, your organization must recognize the need for continuous improvement, as shown in Figure 9–7. Continuous improvement allows a firm to maintain its competitive advantage even when its competitors counterattack.

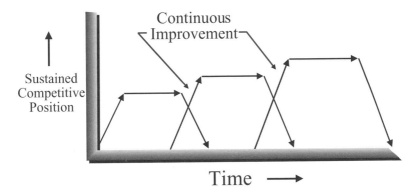

FIGURE 9–7. The need for continuous improvement.

CAPACITY PLANNING

As companies become excellent in project management, the benefits of performing more work in less time and with fewer resources become readily apparent. The question, of course, is how much more work can the organization take on? Companies are now struggling to develop capacity planning models to see how much new work can be undertaken within the existing human and nonhuman constraints.

Figure 9–8 illustrates the classical way that companies perform capacity planning. The approach shown holds true for both project- and non–project-driven organizations. The "planning horizon" line indicates the point in time for capacity planning. The "proposals" line indicates the manpower needed for approved internal projects or a percentage (perhaps as much as 100 percent) for all work expected through competitive bidding. The combination of this line and the "manpower requirements" line, when compared against the current staffing, provides an indication of capacity. This planning technique can be effective if performed early enough such that training time is allowed for future manpower shortages.

There is an important limitation to the above process for capacity planning, however: only human resources are considered. A more realistic method would be to use the strategy shown in Figure 9–9, which can also be applied to both project-driven and non–project-driven organizations. Using the approach shown in Figure 9–9, projects are selected based upon such factors as strategic fit, profitability, consideration of who the customer is, and corporate benefits. The objectives for the projects selected are then defined in both business and technical terms, because there can be both business and technical capacity constraints.

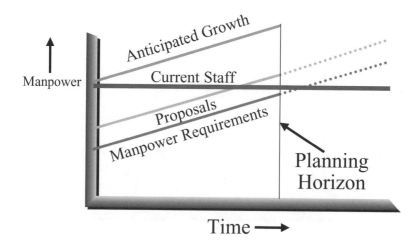

FIGURE 9–8. Capacity planning.

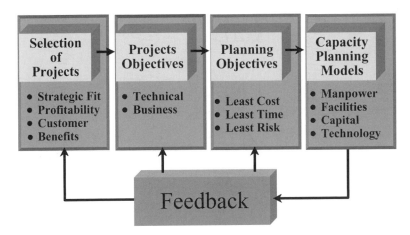

FIGURE 9–9. Capacity planning.

The next step points up one critical difference between average companies and excellent companies. An excellent company will identify capacity constraints from the summation of the schedules and plans. Project managers will meet with project sponsors to determine the objective of the plan, which is different than the objective of the project. Is the objective of the plan to achieve the project's objective with the least cost, least time, or least risk? Typically, only one of these applies, whereas immature organizations believe that all three can be achieved on every project. This, of course, is unrealistic.

The final box in Figure 9–9 is now the determination of the capacity limitations. Previously, we considered only human resource capacity constraints. Now we realize that the critical path of a project can be constrained not only by available manpower but also by time, facilities, cash flow, and even technology. It is possible to have multiple critical paths on a project other than those identified by classical capacity planning. Each of these critical paths provides a different dimension to the capacity planning models, and each of these constraints can lead us to a different capacity limitation. As an example, manpower might limit us to taking on only four additional projects. Based upon available facilities, however, we might only be able to undertake two more projects, and based upon available technology, we might be able to undertake only one new project.

COMPETENCY MODELS

In the twenty-first century, companies will replace job descriptions with competency models. Job descriptions for project management tend to emphasize the deliverables and expectations from the project manager, whereas competency models emphasize the specific skills needed to achieve the deliverables.

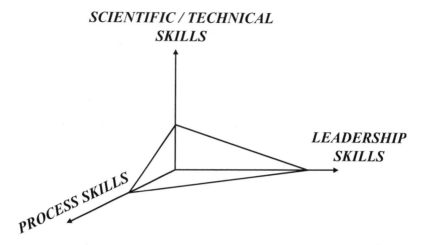

FIGURE 9–10. Competency model for Eli Lilly.

Figure 9–10 shows the competency model for Eli Lilly. Project managers are expected to have competencies in three broad areas:*

- Scientific/technical skills
- Leadership skills
- Process skills

For each of the three broad areas, there are subdivisions or grade levels. A primary advantage of a competency model is that it allows the training department to develop customized project management training programs to satisfy the skill requirements. Without competency models, most training programs are generic rather than customized. Also, competency models make it easier for organizations to develop a complete training curriculum, rather than a single course.

Competency models focus on specialized skills in order to assist project managers in making more efficient utilization of their time. Figure 9–11, although theoretical, shows how, with specialized competency training, project managers might be able to increase their time effectiveness by reducing time robbers and rework. Unfortunately, time robbers and rework cannot always be eliminated, but they can be reduced.

As stated above, competency models make it easier for companies to develop project management curricula, rather than simply single courses. This is shown in Figure 9–12. As companies mature in project management and develop a company-wide core competency model, an internal, custom-designed curriculum will be developed. Companies, especially large ones, will find it necessary to maintain a course architecture specialist on their staff.

---

*A detailed description of the Eli Lilly competency model and the Ericsson competency model can be found in H. Kerzner, *Advanced Project Management.* New York: Wiley, 1999, pp. 266–283.

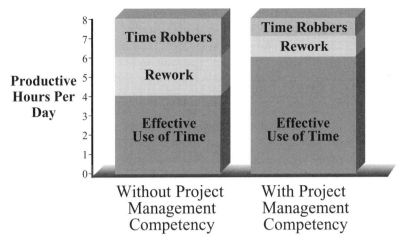

FIGURE 9–11. Core competency analysis.

MANAGING MULTIPLE PROJECTS _____

As organizations begin to mature in project management, there is a tendency to-ward wanting to manage multiple projects. This might entail either the company's sponsoring the various projects, or each project manager's managing multiple projects. There are several factors supporting the managing of multiple projects.

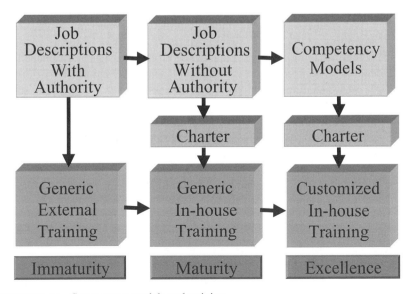

FIGURE 9–12. Competency models and training.

First, the cost of maintaining a full-time project manager on all projects may be prohibitive. The magnitude and risks of the project dictate whether a full-time or part-time assignment is necessary. Assigning a project manager full-time on an activity that does not require it is an overmanagement cost. Overmanagement of projects was considered an acceptable practice in the early days of project management because we had little knowledge on how to handle risk management. Today, methods for risk management exist.

Second, line managers are now sharing accountability with project managers for the successful completion of projects. Project managers are now managing at the template levels of the work breakdown structure (WBS), with the line managers accepting accountability for the work packages at the detailed WBS levels. Project managers now spend more of their time integrating work rather than planning and scheduling functional activities. With the line manager accepting more accountability, time may be available for the project manager to manage multiple projects.

Third, senior management has come to the realization that they must provide high quality training for their project managers if they are to reap the benefits of managing multiple projects. Senior managers must also change the way that they function as sponsors. There are six major areas where the corporation as a whole may have to change in order for the managing of multiple projects to succeed:

- Prioritization: If a project prioritization system is in effect, it must be used correctly such that employee credibility in the system is realized. There are downside risks to a prioritization system. The project manager, having multiple projects to manage, may favor those projects having the highest priorities. It is possible that no prioritization system at all may be the best solution. Also, not every project needs to be prioritized. Prioritization can be a time-consuming effort.

- Scope changes: Managing multiple projects is almost impossible if the sponsors/customers are allowed to make continuous scope changes. When managing multiple projects, the project manager must understand that the majority of the scope changes desired may have to be performed through enhancement projects rather than through a continuous scope change effort on the original projects. A major scope change on one project could limit the project manager's available time to service other projects. Also, continuous scope changes will almost always be accompanied by reprioritization of projects, a further detriment to the management of multiple projects.

- Capacity planning: Organizations that support the management of multiple projects generally have a tight control on resource scheduling. As a precondition, these organizations must have knowledge of capacity planning, theory of constraints, resource leveling, and resource limited planning.

- Project methodology: Methodologies for project management range from rigid policies and procedures to more informal guidelines and checklists. When managing multiple projects, the project manager must be granted some degree of freedom. This necessitates guidelines, checklists, and forms. Formal project management practices create excessive paperwork requirements, thus minimizing the opportunities to manage multiple projects. The project size is also critical.

- Project initiation: Managing multiple projects has been going on for almost 40 years. One thing that we have learned is that it can work well as long the projects are in relatively different life cycle phases. The demands on the project manager's time are different from each life cycle phase. Therefore, for the project manager to effectively balance his/her time among multiple projects, it would be best for the sponsor not to have the projects begin at exactly the same time.
- Organizational structures: If the project manager is to manage multiple projects, then it is highly unlikely that the project manager will be a technical expert in all areas of all projects. Assuming that the accountability is shared with the line managers, the organization will most likely adopt a weak matrix structure.

END-OF-PHASE REVIEW MEETINGS

For more than 20 years, end-of-phase review meetings were simply an opportunity for executives to "rubber-stamp" the project to continue on. The meetings were used to give the executives some degree of comfort concerning project status. Only good news was presented by the project team.

Executives, from a selfish point of view, very rarely cancelled projects. The executive was better off allowing the new product to be developed, even though the executive knew full well that the product would have no buyers or would be overpriced. Once the product was developed, the executive sponsor was "off the hook." The onus now rested on the shoulders of the marketing group to find potential customers. If customers could not be found, obviously the problem was with marketing.

Today, end-of-phase review meetings take on a different dimension. First and foremost, executives are no longer afraid to cancel projects, especially if the objectives have changed, the objectives are unreachable, or if the resources could be used on other activities that have a greater likelihood of success. Executives now spend more time assessing the risks in the future rather than focusing on accomplishments in the past.

Since project managers are now becoming more business-oriented, rather than technically oriented, they are expected to present information on business risks, reassessment of the benefit-to-cost ratio, and any business decisions that could affect the ultimate objectives. Simply stated, the end-of-phase review meetings now focus more on business decisions than on technical decisions.

STRATEGIC SELECTION OF PROJECTS

What a company wants to do is not always what it can do. The critical constraint is normally the availability and quality of the critical resources. Companies usually have an abundance of projects they would like to work on but, because of resource limitations, they have to develop a prioritization system for the selection of projects.

One commonly used selection process is the portfolio classification matrix shown in Figure 9–13. Each potential project undergoes a situational assessment for strengths, weaknesses, opportunities, and threats. The project is then ranked on the nine-square grid, based upon its potential benefits and the quality of resources needed to achieve those benefits. The characteristics of the benefits appear in Figure 9–14, and the characteristics of the resources needed are shown in Figure 9–15.

This classification technique allows for proper selection of projects, as well as providing the organization with the foundation for a capacity planning model to see how much work the organization can take on. Companies usually have little trouble figuring out where to assign the highly talented people. The model, however, provides guidance on how to make the most effective utilization of the average and below average individuals as well.

The boxes in the nine-square grid of Figure 9–13 can then be prioritized according to strategic importance, as shown in Figure 9–16. If resources are limited but funding is adequate, the boxes identified as "high priority" will be addressed first.

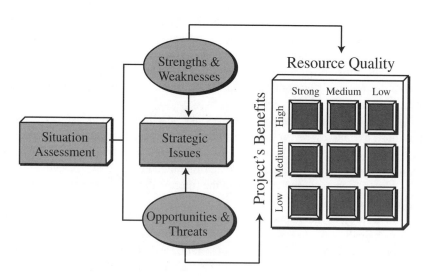

FIGURE 9–13. Portfolio classification matrix.

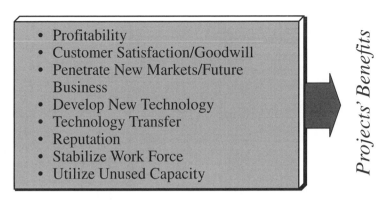

FIGURE 9–14. Potential benefits of a project.

The nine-square grid in Figure 9–16 can also be used to identify the quality of the project management skills needed, in addition to the quality of functional employees. This is shown in Figure 9–17. As an example, the project managers with the best overall skills will be assigned to those projects that are needed to protect the firm's current position. Each of the nine cells in Figure 9–17 can be described as follows:

- Protect position (high benefits and high quality of resources): These projects may be regarded as the survival of the firm. These projects mandate professional project management, possibly certified project managers, and the organization considers project management as a career path posi-

FIGURE 9–15. Characteristics of the resources needed to achieve a project's benefits.

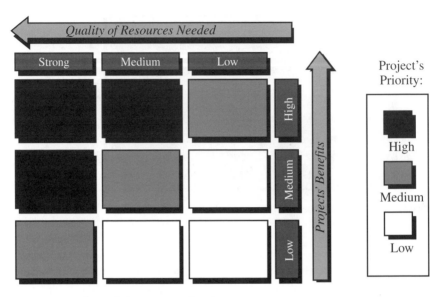

FIGURE 9–16. Strategic importance of projects.

tion. Continuous improvement in project management is essential to make sure that the methodology is the best it can be.

● Protect position (high benefits and medium quality of resources): Projects in this category may require a full-time project manager, but not necessarily a certified one. An enhanced project management methodol-

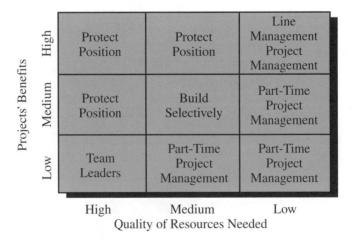

FIGURE 9–17. Strategic guide to allocating project resources.

ogy is needed with emphasis on reinforcing vulnerable areas of project management.

● Protect position (medium benefits and high quality of resources): Emphasis in these projects is on training project managers, with special attention to their leadership skills. The types of projects here are usually efforts to add customer value rather than to develop new products.

● Line management project management (high benefits and low quality of resources): These projects are usually process improvement efforts to support repetitive production. Minimum integration across functional lines is necessary, which allows line managers to function as project managers. These projects are characterized by short time frames.

● Build selectively (medium benefits and medium quality of resources): These projects are specialized, perhaps repetitive, and focus on a specific area of the business. Limited project management strengths are needed. Risk management may be needed, especially technical risk management.

● Team leaders (low benefits but high quality of resources): These are normally small, short-term R&D projects that require strong technical skills. Since minimal integration is required, scientists and technical experts will function as team leaders. Minimal knowledge of project management is needed.

● Part-time project management (medium benefits and low quality of resources): These are small capital projects that require only an introductory knowledge of project management. One project manager could end up managing multiple small projects.

● Part-time project management (low benefits and medium quality of resources): These are internal projects or very small capital projects. These projects have small budgets and perhaps a low to moderate risk.

● Part-time project management (low benefits and low quality of resources): These projects are usually planned by line managers but executed by project coordinators or project expediters.

PORTFOLIO SELECTION OF PROJECTS

Companies that are project-driven organizations must be careful about the type and quantity of projects they work on because of the constraints on available resources. Because timing is often critical, it is not always possible to hire new employees and have them trained quickly enough, or to hire subcontractors, whose skills may well be questionable anyway.

Figure 9–18 shows a typical project portfolio.* Each circle represents a

---

*This type of portfolio was adapted from the life cycle portfolio model used for strategic planning activities.

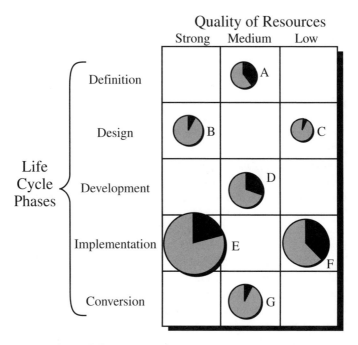

FIGURE 9–18. Basic portfolio.

project. The location of each circle represents the quality of resources needed and the life cycle phase that the project is in. The size of the circle represents the magnitude of the achievable benefits, relative to those of other projects, and the "pie wedge" represents the percentage of the project completed thus far.

In Figure 9–18, Project A has relatively low benefits and uses medium quality of resources. Project A is in the definition phase. However, when Project A moves into the design phase, the quality of resources may change to low or high quality. Therefore, this type of chart has to be updated frequently.

Figures 9–19, 9–20, and 9–21 show three different types of portfolios. Figure 9–19 represents a high-risk project portfolio where high-quality resources are required on each project. This may be representative of a project-driven organization that has been awarded several highly profitable, large projects. This could also be a company that competes in the computer field, an industry that has short product life cycles and where product obsolescence occurs only six months downstream.

Figure 9–20 represents a conservative, profit-oriented project portfolio, say that of an organization that works mainly on low-risk projects that require low-quality resources. This could be representation of project portfolio selection in a service organization, or even a manufacturing firm that has projects designed mostly for product enhancement.

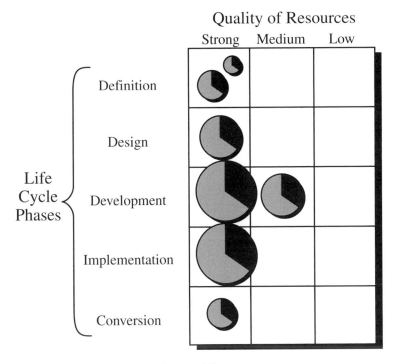

FIGURE 9–19. Typical high-risk project portfolio.

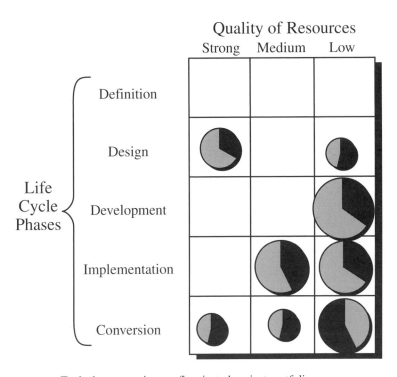

FIGURE 9–20. Typical conservative, profit-oriented project portfolio.

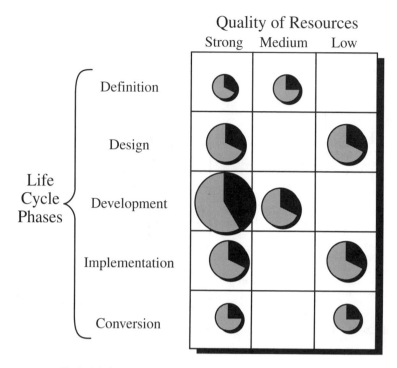

FIGURE 9–21. Typical balanced project portfolio.

Figure 9–21 shows a balanced portfolio with projects in each life cycle phase and where all quality of resources is being utilized, usually quite effectively. A very delicate juggling act is required to maintain this balance.

HORIZONTAL OR PROJECT ACCOUNTING

In the early days of project management, project management was synonymous with scheduling. Project planning meant simply laying out a schedule with very little regard for costs. After all, we know that costs will change (i.e., most likely increase) over the life of the project and that the final cost will never resemble the original budget. Therefore, why worry about cost control?

Recessions and poor economic times have put pressure on the average company to achieve better cost control. Historically, costs were measured on a vertical basis only. This created a problem in that project managers had no knowledge of how many hours were actually being expended in the functional areas to perform the assigned project activities. Standards were very rarely updated and, if they were, it was usually without the project manager's knowledge.

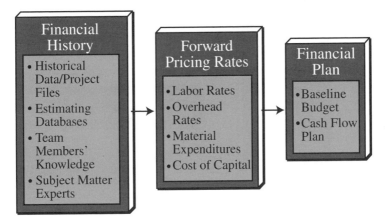

FIGURE 9–22. The evolution of integrated cost-schedule management. Phase I—Budget-based planning.

Today, methodologies for project management mandate horizontal accounting using earned value measurement techniques. This is extremely important, especially if the project manager has the responsibility for profit and loss. Projects are now controlled through a series of charge numbers or cost account codes assigned to all of the work packages in the WBS.

Strategic planning for cost control on projects is a three-phase effort, as shown in Figures 9–22 through 9–24. The three phases are:

- Phase I—Budget-based planning (Figure 9–22): This is the development of a project's baseline budget and cash flow based upon reasonably accurate historical data. The historical databases are updated at the end of each project.
- Phase II—Cost/performance determination (Figure 9–23): This is where the costs are determined for each work package and where the actual costs are compared against the actual performance in order to determine the true project status.
- Phase III—Updating and reporting (Figure 9–24): This is the preparation of the necessary reports for the project team members, line managers, sponsors, and customer. At a minimum, these reports should address the questions of:
  - Where are we today (time and cost)?
  - Where will we end up (time and cost)?
  - What problems do we have now and will we have in the future, and what mitigation strategies have we come up with?

Good methodologies provide the framework for gathering the information to answer these questions.

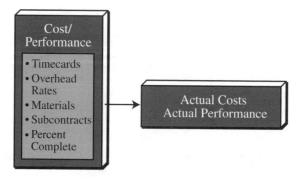

FIGURE 9–23. The evolution of integrated cost-schedule management. Phase II— Cost/performance determination.

ORGANIZATIONAL RESTRUCTURING

Effective project management cultures are based on trust, communication, cooperation, and teamwork. When the basis of project management is strong, organizational structure becomes almost irrelevant. Restructuring an organization only to add project management is unnecessary and perhaps even dangerous. Companies may need to be restructured for other reasons, such as making the customer more important. But successful project management can live within any structure, no matter how awful the structure looks on paper, just as long as the culture of the company promotes teamwork, cooperation, trust, and effective communication.

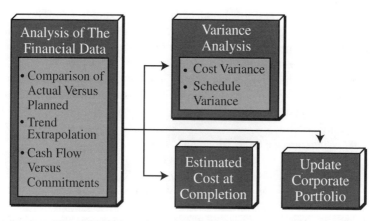

FIGURE 9–24. The evolution of integrated cost-schedule management. Phase III— Updating and reporting.

The organizations of companies excellent in project management can take almost any form. Today, small- to medium-size companies sometimes restructure to pool management resources. Large companies tend to focus on the strategic business unit as the foundation of their structures. Many companies still follow matrix management. Any structure can work with project management as long as it has the following traits:

- The company is organized around nondedicated project teams.
- It has a flat organizational hierarchy.
- It practices informal project management.
- It does not consider the reporting level of project managers to be important.

The first point listed above may be somewhat controversial. Dedicated project teams have been a fact of life since the late 1980s. Although there have been many positive results from dedicated teams, there has also been a tremendous waste of manpower coupled with duplication of equipment, facilities, and technologies. Today, most experienced organizations believe that they are scheduling resources effectively so that multiple projects can make use of scarce resources at the same time. And, they believe, nondedicated project teams can be just as creative as dedicated teams, and perhaps at a lower cost.

Although tall organizational structures with multiple layers of management were the rule when project management came on the scene in the early 1960s, today's organizations tend to be lean and mean, with fewer layers of management than ever. The span of control has been widened, and the results of that change have been mass confusion in some companies but complete success in others. The simple fact is that flat organizations work better. They are characterized by better internal communication, greater cooperation among employees and managers, and atmospheres of trust.

In addition, today's project management organizations, with only a few exceptions (purely project-driven companies), prefer to use informal project management. With formal project management systems, the authority and power of project managers must be documented in writing. Formal project management policies and procedures are required. And documentation is required on the simplest tasks. By contrast, in informal systems, paperwork is minimized. In the future, I believe that even totally project-driven organizations will develop more informal systems.

The reporting level for project managers has fluctuated between top-level and lower-level managers. As a result, some line managers have felt alienated over authority and power disagreements with project managers. In the most successful organizations, the reporting level has stabilized, and project managers and line managers today report at about the same level. Project management simply works better when the managers involved view each other as peers. In large projects, however, project managers may report higher up, sometimes to the executive level. For such projects, a project office is usually set up for project team members at the same level as the line managers with whom they interact daily.

To sum it all up, effective cross-functional communication, cooperation, and trust are bound to generate organizational stability. Let's hope that organizational restructuring on the scale we've seen in recent years will no longer be necessary.

CAREER PLANNING

In organizations that successfully manage their projects, project managers are considered professionals and have distinct job descriptions. Employees traditionally are allowed to climb one of two career ladders: the management ladder or the technical ladder. (They cannot, however, jump back and forth between the two.) This presents a problem to project managers, whose responsibilities bridge the two ladders. To solve this problem, some organizations have created a third ladder, one that fills the gap between technology and management. It is a project management ladder, with the same opportunities for advancement as the other two.

ASSESSMENT INSTRUMENT FOR LEVEL 5

The following 16 questions concern how mature you believe your organization to be with regard to Level 5. Beside each question you will circle the number that corresponds to your opinion. In the example below, your choice would have been "Slightly Agree."

$$-3 \quad \text{Strongly Disagree}$$
$$-2 \quad \text{Disagree}$$
$$-1 \quad \text{Slightly Disagree}$$
$$0 \quad \text{No Opinion}$$
$$+1 \quad \text{Slightly Agree}$$
$$+2 \quad \text{Agree}$$
$$+3 \quad \text{Strongly Agree}$$

Example: $(-3, \quad -2, \quad -1, \quad 0, \quad +1, \quad +2, \quad +3)$

The row of numbers from -3 to $+3$ will be used later for evaluating the results. After answering Question 16, you will grade the exercise by completing Exhibit 5.

QUESTIONS _____

Answer the following questions based upon continuous improvement changes over the past 12 months only. Circle the answer you feel is correct.

1. The improvements to our methodology have pushed us closer to our customers. $(-3 \quad -2 \quad -1 \quad 0 \quad +1 \quad +2 \quad +3)$

2. We have made software enhancements to our methodology. $(-3 \quad -2 \quad -1 \quad 0 \quad +1 \quad +2 \quad +3)$

3. We have made improvements that allowed us to speed up the integration of activities. $(-3 \quad -2 \quad -1 \quad 0 \quad +1 \quad +2 \quad +3)$

4. We have purchased software that allowed us to eliminate some of our reports and documentation. $(-3 \quad -2 \quad -1 \quad 0 \quad +1 \quad +2 \quad +3)$

5. Changes in our training requirements have resulted in changes to our methodology. $(-3 \quad -2 \quad -1 \quad 0 \quad +1 \quad +2 \quad +3)$

6. Changes in our working conditions (i.e., facilities, environment) have allowed us to streamline our methodology (i.e., paperwork reduction). $(-3 \quad -2 \quad -1 \quad 0 \quad +1 \quad +2 \quad +3)$

7. We have made changes to the methodology in order to get corporate-wide acceptance. $(-3 \quad -2 \quad -1 \quad 0 \quad +1 \quad +2 \quad +3)$

8. Changes in organizational behavior have resulted in changes to the methodology. $(-3 \quad -2 \quad -1 \quad 0 \quad +1 \quad +2 \quad +3)$

9. Management support has improved to the point where we now need fewer gates and checkpoints in our methodology. $(-3 \quad -2 \quad -1 \quad 0 \quad +1 \quad +2 \quad +3)$

10. Our culture is a cooperative culture to the point where informal rather than formal project management can be used, and changes have been made to the informal project management system. $(-3 \quad -2 \quad -1 \quad 0 \quad +1 \quad +2 \quad +3)$

11. Changes in power and authority have resulted in looser methodology (i.e., guidelines rather than policies and procedures). $(-3 \quad -2 \quad -1 \quad 0 \quad +1 \quad +2 \quad +3)$

12. Overtime requirements mandated change in our forms and procedures. $(-3 \quad -2 \quad -1 \quad 0 \quad +1 \quad +2 \quad +3)$

13. We have changed the way we communicate with our customers. $(-3 \quad -2 \quad -1 \quad 0 \quad +1 \quad +2 \quad +3)$

14. Because our projects' needs have
changed, so have the capabilities of our
resources. $(-3 \quad -2 \quad -1 \quad 0 \quad +1 \quad +2 \quad +3)$

15. (If your organization has restructured)
Our restructuring caused changes in
signoff requirements in the
methodology. $(-3 \quad -2 \quad -1 \quad 0 \quad +1 \quad +2 \quad +3)$

16. Growth of the company's business base
has caused enhancements to our
methodology. $(-3 \quad -2 \quad -1 \quad 0 \quad +1 \quad +2 \quad +3)$

Exhibit 5

Each response you circled in Questions 1–16 had a column value between -3
and $+3$. In the appropriate spaces below, place the circled value (between -3
and $+3$) beside each question.

1. _____
2. _____
3. _____
4. _____
5. _____
6. _____
7. _____
8. _____
9. _____
10. _____
11. _____
12. _____
13. _____
14. _____
15. _____
16. _____
TOTAL: _____

The grading system for this exercise follows.

EXPLANATION OF POINTS FOR LEVEL 5

Scores 20 or more are indicative of an organization committed to benchmarking
and continuous improvement. These companies are probably leaders in their

field. These companies will always possess more project management knowledge than both their customers and their competitors.

Scores between 10–19 are indicative that some forms of continuous improvement are taking place, but the changes may be occurring slowly. There may be resistance to some of the changes, most likely because of shifts in the power and authority spectrum.

Scores less than 10 imply a strong resistance to change or simply a lack of senior management support for continuous improvement. This most likely occurs in low technology, non–project-driven organizations where projects do not necessarily have a well-defined profit-loss statement. These organizations will eventually change only after pressure by their customers or an erosion of their business base.

Sustainable Competitive Advantage

INTRODUCTION

To spend time and money developing a project management methodology because you believe it is the right thing to do is a wasted effort. The better approach is to develop a methodology with the intent of converting it into a sustainable competitive advantage. A sustainable competitive advantage not only placates your customers, it also puts pressure on your competitors to spend money to compete with you.

Sustainable competitive advantages can be determined for individual functional areas rather than for the entire company. As an example, consider Figure 10–1, which illustrates the efforts needed to achieve a sustained competitive advantage in research and development (R&D). As a company advances through the various stages of innovation, the technical risks will increase. The organization must have developed a good approach to the problem of assessing technical risks and must be willing to admit when a project should be cancelled because the resources could be allocated more effectively on other projects. Maintaining a competitive advantage requires a continuous stream of new and/or enhanced products or services. Risk management is an essential ingredient in the evaluation process.

As technical risks increase, so does the amount of money expended, as well as the requirement for superior technical ability. The technical skills required increase as we go from basic to applied research and on through development. Although some people may argue about the need for this increase in skill levels, the fact remains that a product that can be developed on a small laboratory bench may never be able to be mass-produced or, even if it can be mass-produced, the

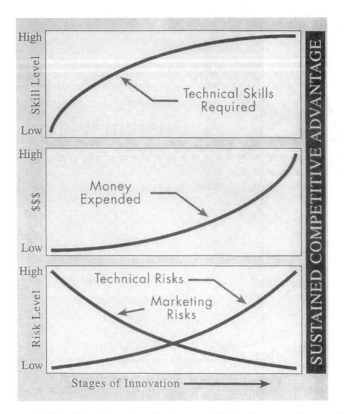

FIGURE 10–1. R&D efforts for a sustained competitive advantage. *Source:* Reprinted from P. Rea and H. Kerzner, *Strategic Planning.* New York: Wiley, 1997, p. 105.

quality may have to be degraded. Also, it is in development where one finally obtains the hard numbers as to whether the product can be manufactured at a competitive price.

STRATEGIC THRUSTS

As shown in Figure 10–2, there are four "strategic thrusts" that must be considered before your project management methodology can be turned into a sustained competitive advantage. These strategic thrusts must be identified while the methodology is being designed and developed, not later on. Developing a methodology and then having to make major changes to it because the strategic thrusts were not considered can waste time and money, as well as lowering morale. Poor morale can cause the workers to lose faith in the methodology.

FIGURE 10–2. Strategic thrusts.

The first strategic thrust is the core values/purpose. The core values/purpose thrust describes the heart of the company, as well as the basic reason for its existence.

- Core values: There are usually three to five core values for a company, the timeless, passionately held guiding principles of the organization. At Procter & Gamble, for example, the core values are delivering consumer value, developing breakthrough innovation, and building strong brands. The core values for the Walt Disney Company might be imagination and wholesomeness, while at Nordstrom they could be service to the customer, trust, and products with style. Core values come from within the organization; they represent what the organization is at its very essence, as opposed to what it does from day to day.
- The core purpose: An organization's core purpose should last for at least 100 years; it is the organization's reason for being that goes beyond current products and services. For 3M, the core purpose is "to solve unsolved problems innovatively." For Hewlett-Packard, it is "to make technical contributions for the advancement and welfare of humanity." For McKinsey & Company, it is "to help leading corporations and govern-

ments to be more successful." For Merck it is "to preserve and improve human life." And for the Walt Disney Company it is "to make people happy." One approach to finding a core purpose is to ask five whys. Start with a description of the business and ask, "Why is that important?" five times; after a few whys you get to the very essence of the business.*

Generally speaking, all projects undertaken using the project management methodology must support the company's core values/purpose, which could very well be regarded as the most important strategic thrust.

The second strategic thrust in Figure 10–2 is the strategic focus. The strategic focus identifies the product/market element in which the organization competes. There are three primary questions that must be addressed in the strategic focus:

- Where will the organization compete? (What products are offered, which markets are served, by segment or geographically?)
- Against whom will the organization compete? (Who is the competition?)
- How will the organization compete? (By product, by proper positioning, by functional strategy as channels of distribution, etc.?)

The answers to these three questions provide guidance on the quality and competencies of the resources and assets needed. Project management methodologies *must* be designed around the competencies of the resources.

The third strategic thrust is the competitive focus. Although this thrust has some similarities to the strategic focus thrust, there are other overriding factors. The competitive focus emphasizes the differences between your organization and your major competitors. The differences can exist in such areas as:

- Product features
- Product design
- Product performance
- Product quality
- Products offered
- Value-added opportunities
- Brand name and image
- Cost reduction opportunities (i.e., experience curves, labor rates)
- Strategic alliances and partnerships

These strategic competitive differences can give your methodology one step up on the competition.

---

*D. A. Aaker, *Strategic Market Management,* 5th ed. New York: Wiley, 1998; p. 28.

FIGURE 10–3. Risks associated with maintaining a sustainable competitive advantage.

The final strategic thrust in Figure 10–2 is synergy. Synergy reflects the organization's ability to perform more work in less time and with fewer resources. Organizational synergy is a measure of how well the employees cooperate with one another. Does the organization have a cooperative or noncooperative culture? Cooperative cultures allow for the design of a flexible methodology that will take advantage of continuous improvement opportunities.

Because market conditions and the environment can change, continuous improvement is necessary to maintain the sustained competitive advantage. Change generates risk that, if not properly analyzed and mitigated, can cause a firm to lose its competitive advantage. The key here is for the competitive advantage to become a sustainable competitive advantage. Typical risks associated with maintaining a sustainable advantage are shown in Figure 10–3.

THE NEED FOR CONTINUOUS IMPROVEMENT

Sustained competitive advantages require continuous improvement for a firm to maintain its strength in the marketplace. Although new products/services are one way, strengthening one's internal position can also be effective if it results in the introduction of new and/or more sophisticated tools that allow a firm to make faster and better decisions. Tools for the future can be classified as follows:

Resource Analysis Tools

- Resource limited planning
- Resource leveling
- Capacity planning
- Multiproject resource analysis

Cost Analysis Tools

- Earned value forecasting
- Variance analysis
- Trend analysis
- Crashing costs

Risk Analysis Tools

- Risk analysis
- Risk quantification
- Lessons learned databases

Forecasting Analysis Tools

- Technology forecasting
- Forward pricing rates
- Escalation factors
- Market analysis

PROJECT MANAGEMENT COMPETITIVENESS

Figure 10–4 shows the process of developing project management competitiveness. These steps are somewhat similar to the steps in the project management maturity model (PMMM). In the first step in Figure 10–4, the organization undergoes project management training, which leads to the development of project management skills. But even with a reasonable skill base, the organization can still be reasonably immature. The project management skill base must be regarded as a company-wide project management competency designed to benefit the entire company.

This is more than simply obtaining the knowledge. It also includes developing a corporate culture that is based upon effective organizational behavior and creating a well-developed project management methodology, accompanied by the proper supporting tools. The tools can be characterized into three areas, as shown in Figure 10–4.

Once the organization recognizes that project management is a core competency, the organization can convert this competency into a sustainable competi-

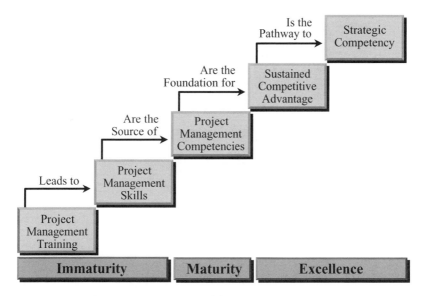

FIGURE 10–4. Project management competitiveness.

tive advantage, as shown in Figure 10–4. The ultimate purpose is for the sustainable competitive advantage to become the pathway for a strategic competency that becomes a primary effort during strategic planning activities. This requires strong executive support and a firm belief that project management does, in fact, impact the bottom line of the corporation.

PRODUCTS VERSUS SOLUTIONS

Quite often the need for strategic planning for project management is driven externally by changing customer requirements and expectations rather than internally. As an example, customers are now requesting that contractors provide them with solutions to their business needs rather than products. When you provide a product to a customer, the customer has the responsibility to unpack the project, inspect it, install it, test it, and get it up and running. When you sell a solution to the customer, then you perform all of those activities for the customer and provide the customer with a totally workable product.

To sell solutions to a customer, you must sell not only the product, but also the delivery system that will achieve the solution. The delivery system is your project management methodology. You must convince the customer that you have the project management methodology to deliver a solution.

Figure 10–5 illustrates how strategic planning for project management now focuses on solutions. Strategic planning for project management has led to the belief that project management now provides the company with a vision that

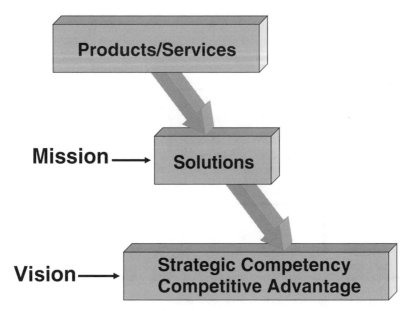

FIGURE 10–5. Identifying the mission and vision.

project management can lead to a competitive advantage and should be regarded as a strategic competency. To achieve this vision, the company establishes an intermediate mission that focuses on providing solutions rather than just products.

Kathy Rhoads, Director, Project Management Center of Excellence at Westfield Group, says:

> Westfield Insurance recognized that the needs of our customers had changed with regard to information technology and information systems projects. Also, the relationship with our customers had changed. Our customers were now expecting complete solutions to their business needs rather than just products. We needed an organization that was designed to partner with our business customers . . .

Achieving this strategic competency and competitive advantage requires that some foundation elements for project management exist. This is shown in Table 10–1, which identifies what must be done in the near term and long term.

ENTERPRISE PROJECT MANAGEMENT

For almost 30 years, project management has meant different things to different people. The focus was always on the end result. Today, the focus is on the delivery system. Historically, every organizational unit within a company was allowed to have its own methodology for producing components, products, and services.

TABLE 10–1. THE FOUNDATION ELEMENTS

Long Term	Short Term
• Mission	• Primary and secondary processes
• Results	• Methodology
• Logistics	• Globalization rollout
• Structure	• Business case development
• Accountability	• Tools
• Direction	• Infrastructure
• Trust	
• Teamwork	
• Culture	

But today, customers want complete solutions where a solution can be defined as the integration of multiple components, products, and services. The result is a requirement for a single methodology, which is called enterprise project management and which spans the entire company. To illustrate this point, consider what took place at Rockwell Automation. Rockwell Automation had multiple functional project management methodologies in place. Each functional division had its own approach for managing projects. Management recognized that Rockwell Automation needed to change. First, every activity at Rockwell Automation was regarded as a project. Therefore, Rockwell Automation was managing their business by projects! Second, their customers were now asking Rockwell Automation for solutions rather than products. The necessity for an enterprise project management methodology was quite apparent.

Figure 10–6 illustrates the evolutionary process that companies go through to bring in enterprise project management.

The vertical arrows reflect functional project management where each functional area, such as marketing or engineering, could have its own form of project management. This is acceptable as long as the customers want products or components. But when a customer wants complete solutions, then all of the functional areas must work together, thus necessitating a single enterprise project management methodology. As companies begin to realize that project management has become a strategic competency, there is a desire for the relationship with customers and suppliers to be one of partnerships rather than just a customer-contractor relationship. Therefore, strategic planning on the enterprise project management methodology must occur such that your methodology can easily adapt itself and interface with the methodologies of your customers and suppliers. This is the meaning of the bottom portion of Figure 10–6.

ENGAGEMENT PROJECT MANAGEMENT

One of the newest terms to enter the project management vocabulary is "engagement project management." Engagement project management involves the way

Functional Project Management (Products)

Enterprise Project Management (Solutions)

Suppliers   **Customers**

Extended Enterprise Project Mgt. (Partnerships)

FIGURE 10–6. Growth of enterprise project management.

you approach a prospective client and how you present what you are selling. Given the fact that clients are looking for strategic, long-term partnerships, as well as for solutions rather than just products, you must sell your project management skills to win future business. To provide solutions and a meaningful partnership, engagement project management must emphasize:

- Realistic deliverables and constraints
- An enterprise project management methodology that can deliver continuously and with a high probability of success
- Continuous status reporting
- Customer involvement during decision-making

Engagement project management has created an understanding and demand for project management in both the buyer and seller's organization.

Special Problems with Strategic Planning for Project Management

INTRODUCTION

Even with strategic planning for project management, special problems can always occur, problems that end up creating difficulties for both the project manager and the organization. Not all situations can be anticipated, of course, but some that might have been anticipated, are often simply not considered. Some problems exist purely because of misconceptions.

Perhaps the most common misconception arises when the company believes that the implementation of a singular methodology is a cure-all for all ailments. Project management is *not* a guarantee of success, but it can drastically improve your chances for success.

The remaining sections in this chapter contain special problems and misconceptions that have recently occurred even in the best-managed companies and also in those companies with world-class methodologies for project management. If the project achieves only 86 percent of the targeted specification, is the project a failure? If the project is canceled because we were researching in the wrong area, is this a success or a failure? If the customer's deliverable was not achieved but the customer was delighted with the working relationship, is the project a success or a failure? Companies today are redefining the meanings of success and failure.

Change management is another topic that must be considered, especially if strategic planning for project management requires people to change they way they had worked for the past several years. Change management is now a top priority for companies that wish to accelerate the project management learning curve. And finally, although this book is not intended to be a course on risk management, several risk management problems are now surfacing that affect the way we perform strategic planning for project management and impact on how we design a singular methodology. Proper design of a singular methodology for project management can simplify the way risk management planning is accomplished, and even encourage the project managers to increase their personal tolerance level for risk.

THE MANY FACES OF SUCCESS

Historically, success has been defined as meeting the customer's expectations, regardless of whether "the customer" was internal or external. Furthermore, success has also meant getting the job done within the constraints of time, cost, and quality. Using this standard definition, success could be visualized as a singular point on a time, cost, quality/performance grid. But how many projects, especially those requiring innovation, can possibly reach this exact point?

Very few projects are ever completed without tradeoffs or scope changes on time, cost, and quality. Therefore, success might still occur without exactly hitting this singular point. In this regard, success might be better defined as a cube, such as seen in Figure 11–1. The singular point of time, cost, and quality would be a point within the cube.

Another factor to consider is that there may exist both primary and secondary definitions of success, as shown in Table 11–1. The primary definitions of success

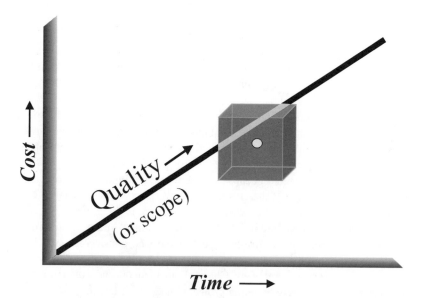

FIGURE 11–1. Success: point or cube?

TABLE 11–1. SUCCESS FACTORS

Primary	Secondary
• Within time • Within cost • Within quality limits • Accepted by the customer	• Follow-on work from this customer • Using the customer's name as a reference on your literature • With minimum or mutually agreed upon scope changes • Without disturbing the main flow of work • Without changing the corporate culture • Without violating safety requirements • Providing efficiency and effectiveness of operations • Satisfying OSHA/EPA requirements • Maintaining ethical conduct • Providing a strategic alignment • Maintaining a corporate reputation • Maintaining regulatory agency relations

are seen through the eyes of the customer. The secondary definitions of success are usually internal benefits. If achieving 86 percent of the specification is acceptable to the customer and follow-on work is received, then the original project might very well be considered as a success.

It is possible for a project management methodology to identify primary and secondary success factors. This could provide guidance to a project manager for the development of a risk management plan, as well as help the project manager decide which risks are worth taking and which are not acceptable.

THE MANY FACES OF FAILURE*

Previously we stated that success might be a cube rather than a point. If we stay within the cube but miss the point, is that a failure? Probably not! The true definition of failure is when the final results are not what were expected, even though the original expectations may or may not have been reasonable. Sometimes customers, and even internal executives, set performance targets that are totally unrealistic; they may hope to achieve 80 to 90 percent at best. For simplicity's sake, let us define failure as unmet expectations.

When unmeetable expectations are formed, failure is virtually assured, since we have defined failure as unmet expectations. This is called a *planning failure* and is the difference between what was planned to be accomplished and what was, in fact, achievable. The second component of failure is poor performance or

---

* Adapted from Robert D. Gilbreath, *Winning at Project Management.* New York: Wiley, 1986, pp. 2–6.

actual failure. This is the difference between what was achievable and what was actually accomplished.

Perceived failure is the net sum of *actual failure* and *planning failure.* Figures 11–2 and 11–3 illustrate the components of perceived failure. In Figure 11-2, *project management* has planned a level of accomplishment (C) lower than what is achievable given project circumstances and resources (D). This is a classic underplanning situation. Actual accomplishment (B), however, was even less than planned.

A slightly different case is illustrated in Figure 11–3. Here, management has planned to accomplish more than can be achieved. Planning failure is again assured even if no actual failure occurs. In both of these situations (overplanning and underplanning), the actual failure is the same, but the perceived failure can vary considerably.

Today, most project management practitioners focus on *planning failure.* If this aspect of the project can be compressed, or even eliminated, then the magnitude of the actual failure, should it occur, would be diminished. A good project management methodology helps to reduce planning failure. Today, we believe that planning failure, when it occurs, is due in large part to the project manager's inability to perform effective risk management. In the 1980s, we believed that the failure of a project was largely a quantitative failure due to:

- Ineffective planning
- Ineffective scheduling
- Ineffective estimating
- Ineffective cost control
- Project objectives being a "moving target"

During the 1990s, we changed our view of failure from being quantitatively oriented to qualitatively oriented. A failure in the 1990s was largely attributed to:

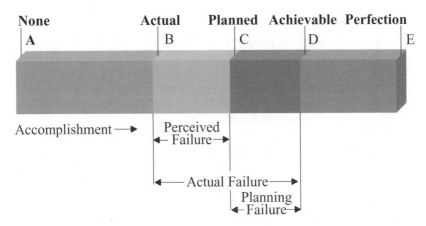

FIGURE 11–2. Components of failure.

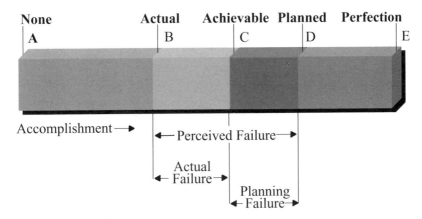

FIGURE 11–3. Components of failure.

- Poor morale
- Poor motivation
- Poor human relations
- Poor productivity
- No employee commitment
- No functional commitment
- Delays in problem-solving
- Too many unresolved policy issues
- Conflicting priorities between executives, line managers, and project managers

Although these quantitative and qualitative approaches still hold true to some degree, today we believe that the major component of planning failure is inappropriate or inadequate risk management, or having a project management methodology that does not provide any guidance for risk management.

Sometimes, the risk management component of failure is not readily identified. For example, look at Figure 11–4. The actual performance delivered by the contractor was significantly less than the customer's expectations. Is the difference between these two arrows poor technical ability or a combination of technical inability and poor risk management? Today we believe that it is a combination.

When a project reaches completion, the company performs a lessons learned review (or at least a well-managed company does). Sometimes lessons learned are inappropriately labeled, and thus the true reason for the risk event remains unknown. Figure 11–5 illustrates the relationship between the marketing personnel and technical personnel when undertaking a project to develop a new product. If the project is completed with actual performance being less than customer expectations, is it because of poor risk management by the technical assessment and

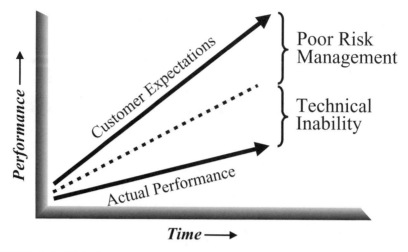

FIGURE 11–4. Risk planning.

forecasting personnel or poor marketing risk assessment? The interrelationship between marketing and technical risk management is not always clear.

Another point illustrated in Figure 11–5 is that opportunities for tradeoffs diminish as we get further downstream on the project. There are numerous opportunities for tradeoffs prior to establishing the final objectives for the project, but more limited chances during project execution. Thus, if the project fails, it may very well be because of the timing when the risks were analyzed.

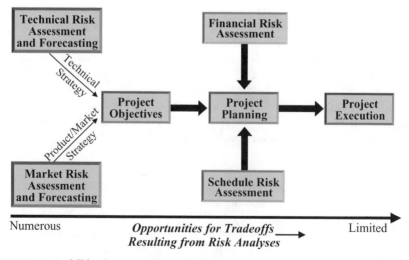

FIGURE 11–5. Mitigation strategies available.

Care must be taken that the proper identification of failure is made and that the company knows to which functional areas the lessons learned should be applied.

TRAINING AND EDUCATION

Given the fact that most companies use the same basic tools as part of their methodology, what then makes one company better than another? The answer lies in the execution of the methodology. Training and education can accelerate not only the project management maturity process but also the ability to execute the methodology.

Actual learning takes place in three areas, as shown in Figure 11–6: on-the-job experience, education, and knowledge transfer. Ideal project management knowledge would be obtained by allowing each employee to be educated on the results of the company's lessons learned studies including risk management, benchmarking, and continuous improvement efforts. Unfortunately, this is rarely done and ideal learning is hardly ever reached. To make matters worse, actual learning is less than most people believe because of lost knowledge. This lost knowledge is shown in Figure 11–7 and will occur even in companies that maintain low employee turnover ratios.

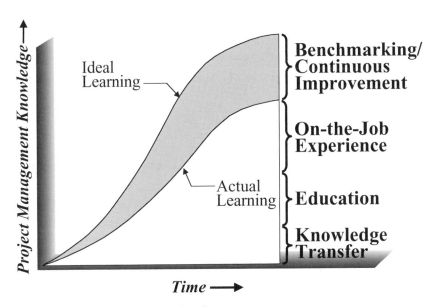

FIGURE 11–6. Project management learning curve.

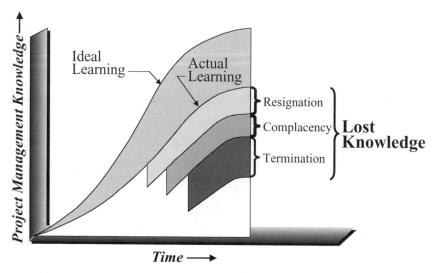

FIGURE 11–7. Project management learning curve.

CULTURAL CHANGE MANAGEMENT

It has often been said that the most difficult projects to manage are those that involve the management of change. Figure 11–8 shows the four basic inputs needed to develop a project management methodology. Each of these four inputs has a "human" side that may require that people change.

Successful development and implementation of a project management methodology requires:

- Identification of the most common reasons for change in project management and why these reasons occur.
- Identification of the ways to overcome the resistance to change.
- Application of the principles of change management to ensure that the desired project management environment will be created and sustained.

For simplicity's sake, resistance can be classified as departmental resistance and personal resistance to change. Organizational resistance occurs when a functional unit as a whole feels threatened by project management. This is shown in Figure 11–9. Examples include:

- **Sales:** The sales staff's greatest resistance to change arises from fear that project management will now take credit for corporate profits, thus reducing the year-end bonuses for the salesforce. Sales personnel fear that

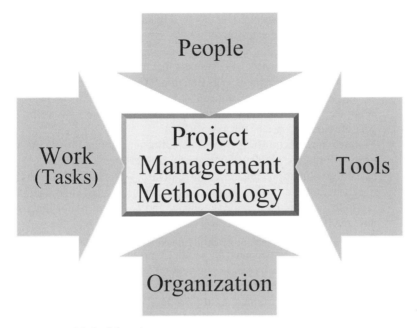

FIGURE 11–8. Methodology imputs.

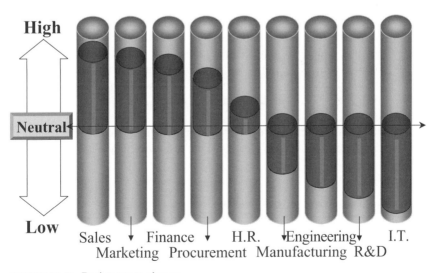

FIGURE 11–9. Resistance to change.

project managers may become involved in the sales effort, thus diminishing the power of the salesforce.

- **Marketing:** Marketing people fear that project managers will end up working so closely with the customers that they may eventually be given some of the marketing and sales functions. This fear is not without merit because customers often want to communicate with the personnel managing the project rather than with marketing staff, who may well disappear after the closure of the sale.

- **Finance (and accounting):** These departments fear that project management will require the development of a project accounting system (such as earned value measurement), which, in turn, will increase the workload in accounting and finance. Accounting personnel are not happy with having to learn earned value measurement, and having to perform accounting both horizontally (i.e., in projects) and vertically (i.e., in line groups).

- **Procurement:** The fear among this group is that a project procurement system will be implemented in parallel with the corporate procurement system, and that the project managers will perform their own procurement, thus bypassing the procurement department.

- **Human resources management:** The H. R. department may fear that a project management career path ladder will be created, not to mention a need for project management training programs. Human resources personnel may find that their traditional training courses, with which they feel very comfortable, have to undergo major change to satisfy project management requirements. This will increase their workloads.

- **Manufacturing:** Little resistance is usually found here because, although the manufacturing segment of a company is not project-driven, staff there are usually familiar with numerous capital installation and maintenance projects, which will have required the use of project management.

- **Engineering, R&D, and Information Technology:** These departments are almost entirely project-driven with very little resistance to project management. Project management is viewed as a necessity.

Getting the support of and partnership with functional management in the involved departments can usually overcome the resistance that exists on an organi-

TABLE 11–2. RESISTANCE: WORK HABITS

Causes of Resistance	Ways to Overcome Resistance
• New guidelines/processes • Need to share "power" information • Creating a fragmented work environment • Giving up established work patterns • Changing comfort zones	• Dictate mandatory conformance from above • Create new comfort zones at an acceptable pace • Identify tangible/intangible individual benefits

TABLE 11–3. RESISTANCE: SOCIAL GROUPS

Causes of Resistance	Ways to Overcome Resistance
• Unknown new relationships • Multiple-boss reporting • Multiple temporary assignments • Severing established ties	• Maintain existing relationships in effect • Avoid cultural shock • Find an acceptable pace for rate of change

zational level. However, the resistance that may occur on an individual level is usually more complex and more difficult to overcome. Individual resistance levels may result from:

● Potential changes in work habits
● Potential changes in the social groups
● Embedded fears
● Potential changes in the wage and salary administration program

Tables 11–2 through 11–5 show how these different issues can lead to resistance, and list possible solutions. It is imperative that we understand the resistance to change. If individuals are happy with their current environment, there will always be resistance to change. But even if people are unhappy with their present environment, there will still exist resistance to change unless: (1) people believe that the change is possible, and (2) people believe that they will somehow benefit from the change. People are usually apprehensive over what they must give up rather than excited about what they will get in return. Not all people will be at the same readiness level for change even if the entire organization is going through the change. People will handle only so much change at one time and, if management eases up on its pressure for change, employees will revert back to their old ways.

Management is the architect for the change process. Management must develop the appropriate strategies such that the organization can align itself with change. This is best done by developing a shared understanding with the employees expected to make the change.

TABLE 11–4. RESISTANCE: EMBEDDED FEARS

Causes of Resistance	Ways to Overcome Resistance
• Fear of failure • Fear of termination • Fear of added workload • Fear of uncertainty/unknowns • Dislike of uncertainty/unknowns • Fear of embarrassment • Fear of a "we/they" organization	• Educate employees on benefits of changes to the individual/corporation • Show a willingness to admit/accept mistakes • Show a willingness to pitch in • Transform unknowns into opportunities • Share information

TABLE 11–5.   RESISTANCE: WAGE AND SALARY ADMINISTRATION

Causes of Resistance	Ways to Overcome Resistance
• Shifts in authority and power • Lack of recognition after the changes • Unknown rewards and punishment • Improper evaluation of personal performance • Multiple-boss reporting	• Link incentives to change • Identify future advancement opportunities/career paths

Performing the following can facilitate the change process:

- Management must explain to the employees the reasons for the change and solicit feedback.
- Management must explain to the employees what outcomes are desired and why.
- Management must champion the change process.
- Management must empower the appropriate individuals to institutionalize the changes.
- Management must be willing to invest in training necessary to support the changes.

Facilitation of the corporate culture change management process may be a necessity when implementing the project management maturity model (PMMM). Project management cultures are the systems, methodologies, assumptions, and conventions that govern the actions of the employees when working on projects. Tools and software do not manage projects. Instead, people, who inevitably have the responsibility to make any and all critical decisions, manage projects. The way people interact, get the work accomplished, and make decisions is dictated by the project management culture.

Corporate cultures may take a long time to create and put into place, but can be torn down overnight. Corporate cultures for project management are based upon organizational behavior, not processes. Corporate cultures reflect the goals, beliefs, and aspirations of senior management. It may take years for the building blocks to be put in place for a good culture to exist, but the whole structure of the culture can be torn down quickly through the personal whims of one executive who refuses to support project management.

Project management cultures can exist within any organizational structure. The speed at which the culture matures, however, may be based upon the size of the company, the size and nature of the projects, and the type of customer, whether it be internal or external. Project management is a culture, not a set of policies and procedures. As a result, it may not be possible to benchmark a project management culture. What works well in one company may not work equally well in another.

Where they exist, however, good corporate cultures can also foster better relations with the customer, especially with external clients. As an example, one company developed a culture of always being honest in reporting the results of testing being performed for external customers. The customers, in turn, began treating the contractor as a partner and routinely shared proprietary information so that the customers and the contractor could help each other out.

Within the excellent companies, the process of project management evolves into a behavioral culture based upon multiple-boss reporting. The significance of multiple-boss reporting cannot be understated. There is a mistaken belief that project management can be benchmarked from one company to another. Benchmarking is the process of continuously comparing and measuring against an organization anywhere in the world in order to gain information that will help your organization improve its performance and competitive position. Competitive benchmarking is the term used when one organization benchmarks its organizational performance against the performance of competing organizations. Process benchmarking, on the other hand, is the benchmarking of discrete processes against organizations with performance leadership in these processes.

Since a project management culture is a behavioral culture, benchmarking works best if we focus our benchmarking on best practices that include leadership, management, or operational methods that lead to superior performance. Because of the strong behavioral influence, it is almost impossible to transpose a project management culture from one company to another. What works well in one company may not be appropriate or cost-effective in another company.

When trying to improve project management, senior management often overemphasizes the quantitative components of the PMMM and underemphasizes the behavioral component. These mistakes are quite common even in the best management companies.

The first step in changing a culture is to identify the key players that can influence the cultural change. In the following situation, the president identified the line managers as a critical change element.

Situation 1: The Automotive Supplier

The Automotive Services Group (ASG) of a large corporation was floundering. Corporate management had great expectations for ASG considering the quality of their technical resources. Performance was significantly below expectations, however, and corporate executives believed that a change in leadership was necessary for project management performance to improve.

The company hired a new divisional president for the ASG. After interviewing personnel at all levels of the organization, the president came to believe that the root cause of the problem was the culture at ASG. Cooperation was poor at best, and technical decision-making was dominated by the line managers. In many cases, less than optimal decisions were made and then implemented. Project management, simply stated, was not working well nor was it providing the best results.

To create a cooperative corporate culture, the new president's attention focused first on the line managers. The president believed that line managers had a much greater influence on the culture than did project managers. The president discovered that most of the line managers were technical experts who had been promoted to management because they were at the top of the pay grades for their group and could not receive additional compensation without becoming a line manager. Many of these people had not really wanted to become line managers but had accepted the position reluctantly for the additional income.

The president was granted complete autonomy by the board of directors to implement whatever changes were necessary to turn around the ASG. The four-step approach that was implemented included:

- Creation of a new pay grade called technical consultant whereby management could reward through almost unlimited salary highly technical employees without forcing them to become line managers.
- Giving several of the existing technical line managers the option of relinquishing their line management position and opting instead for the new consultant pay scale.
- Expecting line managers to function more as administrative managers than as technical managers. Also, expecting line managers to have an understanding of technology but not necessarily a command of technology.
- Establishing that the most important criteria for promotion to line manager would be a demonstrated willingness to cooperate with other line managers and project managers, as well as willingness to make decisions in the best interest of the company rather than the line function or the project.

The president focused heavily on the importance of cooperation. For more than six months, the president conducted personal interviews with employees and managers to see if a cooperative culture was being developed. Within two years the company developed a project management methodology that was being implemented more informally than formally because of the cooperative culture that had developed. Performance improved significantly.

Situation 2: The Aerospace Company

A line manager in an aerospace firm had the responsibility to provide deliverables to four high priority projects. These were the only projects labeled as high priority. The line manager needed at least 20 employees from his group to provide all of the deliverables on time. Unfortunately, the line manager had only 12 people available. The line manager could have simply made the decision according to the established priorities as to which projects would be serviced first. Another option would have been to elevate the problem to senior management who most likely would also follow the established priorities.

Because the organization had a cooperative culture, however, the line manager instead asked the four project managers for help. All four project managers sat around the conference table, exposed their contingency plans to one another, and made tradeoffs such that none of the four project managers would be significantly hurt. The four project managers and the line manager then informed senior management of their recommendation. Senior management's response was, "Thank you! Goodbye!"

Another common mistake made by senior management is believing that project management cultures can be transplanted from one division of a business to another or from one company to another. Organizational change fails when culture is ignored. Project management cultures are based upon the size and nature of the projects, the type of customers, and the requirement of the projects. In the next situation, an executive mistakenly believed that there was very little difference between aerospace projects and consumer projects.

Situation 3: The Commercial Products Division

The Commercial Products Division (CPD) of a large company identified a need for project management. With very little guidance on how to change the existing culture to a project management culture, the company transferred two of their senior managers from their Aerospace Division to the Commercial Products Division. Both managers had spent at least 25 years in the Aerospace Division. The intent was to create the same culture that existed in the project-driven Aerospace Division in the non-project-driven CPD.

The two newly transferred managers erroneously believed that cultural transfers would be easy to accomplish. Project management software that was being used on large aerospace projects was installed on the desktop computers of the project managers and line managers in the CPD. Forms, policies, procedures, and templates that were used on aerospace projects were also installed as part of the new CPD methodology for project management. The CPD project management methodology was almost identical to the aerospace division's project management methodology.

It took less than two years for the company to realize that a mistake had been made. The two managers returned to the Aerospace Division for reassignment. The company instructed the CPD to develop its own methodology and culture for project management. But the damage had been done, and was perhaps even irrevocable. Significant resistance to project management had now developed. The scars left on the corporate culture would take time to repair.

When a significant change occurs to the business base of a company, a corresponding change to the corporate culture may be needed. This cultural change could be for the better or for the worse. When executives make a decision to change the strategic direction of the company, they should assess the potential impact on the culture. If this is not done, then irrevocable damage can occur. This point is illustrated in the next two situations.

Situation 4: The Construction Company

Davis Construction Company had divisions in several large U.S. cities. Davis had a reputation for performing quality work and usually bid on only small- to medium-sized jobs. Small jobs were considered to be under $10 million and medium-sized jobs were $10–$15 million. Occasionally Davis would bid on a large job, but most of the work then had to be subcontracted out to selected partners.

Senior management at Davis was relatively conservative. The company wanted no more than a 3 to 5 percent growth rate per year in sales. The revenue stream was fairly stable, and the firm's financial health was good.

When the U.S. economy began to derail in the early 1990s, Wall Street analysts were predicting that gasoline prices at the pumps could exceed $2 per gallon by year end. If prices reached or exceeded $2 per gallon, the belief was that companies would invest heavily in alternate energy sources, synthetic fuel plants, oil shale research and production, and possibly a resurrection of nuclear power plant construction. These projects could easily reach $400–$500 million in size, with nuclear power plants at about $8 billion or more.

Senior management at Davis believed that this was a once-in-a-lifetime opportunity to compete head on with the larger construction companies and to reap large profits. The decision was thus made to focus heavily on the larger projects, over $100 million in size, and to give up many of the smaller projects that could tie up critical resources.

The existing project management methodology had been designed to service small projects and could not effectively support mega-projects. The company considered having two separate methodologies: one for small projects and another one for large projects. Since the company considered its future existence to be heavily dependent on large projects, a new methodology was created for large projects and the old methodology was discarded.

The company began bidding on the larger projects. Each time a large project would be awarded to Davis, hundreds of engineers would be hired. Within two years, the company quadrupled in size and the business was flourishing. The future looked promising.

Nobody seemed to notice or care that the culture at Davis had changed. With the old methodology, emphasis had been on teamwork and cooperation. With the new methodology, emphasis was on policies and procedures. Everything was done according to policies and procedures for fear that mistakes could lead to massive cost overruns. The culture had changed at the expense of profitability.

As the economy began to improve, it became evident that gas prices would fall to approximately $1 per gallon. Companies that had prepared plans for mega-projects put everything on hold. Projects that had already been started were being canceled.

For Davis Construction, these changes were disastrous. Mega-projects were needed to support the enlarged resource base. But with fewer mega-projects available and competition becoming fierce, downsizing was inevitable. Davis tried vehemently to change back to the old methodology, but this was an impossible task during downsizing when people saw their jobs disappearing. Survival and internal competition took on paramount importance for the workers. Davis is no longer in business.

Situation 5: The Software Solutions Group (SSG)

Three graduate students who had completed a graduate program in information systems started up SSG in the early 1990s. By 2001, the company employed over 250 people servicing exclusively U.S. based companies. Unfortunately, the downturn in technology spending was taking its toll on SSG's profitability. SSG had to either downsize its organization or find additional work.

SSG's core business was supported by some 30 large companies that found it less expensive to outsource work to SSG than to perform the same software development efforts internally. SSG assigned their best systems programmers to these 30 clients. The working relationships with the 30 clients were on solid ground, and the revenue stream from these companies was stable. Unfortunately, this revenue stream could not support the entire organization.

To maintain cash flow, the company began bidding aggressively for work outside of the United States. Many of the contacts were priced out just to meet expenditures and salaries and keep the present workforce employed. The company's business base increased, but problems began to surface.

Because many of the new global customers were in different time zones, some with as much as six hours difference, the company went to flex-time. Workers were allowed to start and finish work whenever they wanted as long as they still put in an eight-hour day. This alleviated some of the problems with time zone changes, but it also began to change the social aspect of the culture within SSG.

In order to win contracts, SSG accepted statements of work that had ill-defined requirements. Many of SSG's new clients were clueless as to what was meant by a well-defined statement of work (SOW). The number of scope changes was increasing exponentially. The project managers working on multinational projects began "stealing" resources from other projects to put out fires on their own projects. This caused havoc with the projects underway for SSG's 30 core customers, who could no longer contact the resources who had been assigned to their projects previously. Maintaining the schedules on the multinational projects was being accomplished at the expense of the domestic business base. And, even more unfortunately, the number of scope changes was expected to increase.

The president recognized that the cooperative culture that had taken ten years to build and nurture was becoming derailed rapidly. The president immediately restructured the company on the mistaken belief that organizational structure shapes culture. To make matters worse, the president then put the project managers on salary and commission rather than just salary. The commission was a percentage of project profits.

As the number of scope changes grew, it became apparent that the multinational customers' needs were unstable. The stealing and hoarding of precious resources began to increase. The organizational culture simply could not cope with the constantly changing requirements and the new compensation system. Quality resources began leaving the company for employment elsewhere. The corporate culture had disintegrated. The president realized that he was better off with a small, focused customer base rather than marketing to the world. But the damage was done and he was unsure as to how to correct the situation quickly.

Changing a corporate culture to one of cooperation and teamwork can be viewed as a project by itself. Unfortunately, as seen in the next situation, not everyone will accept the new culture.

Situation 6: The Junior High School Boys Basketball Team

During preseason practice games with other junior high schools, the coach became concerned that the "culture" of the team was so poor that a losing season was inevitable. Teamwork was simply not there. Whoever had the ball would shoot, even when someone else was closer to the basket and would have an easier shot. Each person played as an individual, looking for points rather than assists.

The coach came up with an interesting idea. During the next team practice, the coach placed boxing gloves in the middle of the basketball court. Each player had to place one boxing glove on either of his hands. The rules of the game were that if you were defending a player holding the basketball, you could "bash" him in the arm, chest, or stomach as long as he held the basketball. The intent was to force the players to pass the ball and play as a team.

Each of the players selected a boxing glove for the hand that could deliver the most powerful punch. Right-handed players put on a right-handed glove. Left-handed players put on a left-handed glove. As the practice game progressed, significant bashing took place, but very few points were scored. The reason for the few points was that the strong hand that had the boxing glove was also the hand used for shooting the basketball.

At the next practice session, most of the players chose a boxing glove for the non-shooting hand. Teamwork improved somewhat, and more points were being scored. However, there were still some players who selected the boxing glove for the strong hand and preferred bashing rather than winning games and teamwork.

Corporate cultures are inevitably impacted by the wage and salary administration program. Care must be taken that whatever changes are required to meet the future needs of the business do not have a major impact on the wage and salary administration program. If the impact is not fully understood and mishandled, irrevocable damage can occur.

Situation 7: The Electronics Manufacturer

Several years ago, a manufacturer of electronic components recognized the need to implement project management. A new department for project managers was created. Since project managers were expected to work closely with the department managers for the accomplishment of deliverables, the salary structure of the project managers was close to that of the line managers. The president believed that the working relationship between the project and line managers would function best if both parties viewed each other as equals rather than in a superior-subordinate relationship.

The corporate culture liked what the president had instituted. Over a period of several years, the culture fostered an atmosphere of trust and effective cooperation. In 2002, the president retired and a new CEO was brought in.

The new president liked the way that project management was working but believed that it could be working even better. Project managers were offered a financial incentive, similar to a bonus, based upon the profitability of their projects. Only the project managers were offered the incentives.

Project managers began to focus more on the incentives than on the working relationship with line managers. To make matters worse, several project managers were now receiving total compensation packages that were significantly larger than their functional counterparts.

The relationship between the project and line managers began to deteriorate, and this, of course, had a significant detrimental impact on the corporate culture. The president began filing vacant line management positions with project managers who had been successful in project management.

Employees began believing that the only career path to the top would be through project management. Employees and line managers began volunteering to become project managers. The technical base of the company was becoming diluted because everyone began to believe that "the grass was greener" in project management than in line management. The corporate culture for executing projects was now focusing on animosity and jealousy rather than cooperation and teamwork.

Perhaps the greatest challenge facing senior managers in this decade will be the creation of a multinational project management culture. Cultural change requires transformations to take place, and in some countries these transformations are difficult to implement. In the next situation, the resistance to change occurred at the executive levels.

Situation 8: The Global Automotive Supplier

A United States based automotive supplier had developed an excellent project management methodology. All of the U.S. divisions used the methodology with outstanding results. Their customers liked the methodology and the quality of the deliverables to such a degree that they began treating the company as a partner rather than as an ordinary supplier.

As business began to grow, the company began to purchase other companies around the world, planning to become a global automotive supplier. The original intent was to leave local management in place and to allow each division to operate autonomously as long as business was good. However, this meant that multiple project management methodologies and cultures would be in existence. Some of the newly acquired divisions had no project management in place at all, with all projects being managed through line managers. Other divisions had project management but functioned without the use of a project management methodology.

Realizing that there would be a rapid growth in the number of projects requiring several divisions to work together, the company decided it needed to have all divisions managing projects the same way. Project managers and professional trainers on project management were brought from the United States to each division to help install the project management methodology and to change the culture, if necessary.

Strong resistance to project management became evident at all levels of management in the newly acquired divisions, especially at the executive levels. One executive stated that as long as the customers in his country did not recognize project management as being as important as the American auto industry, he would not support project management.

After two years of battling the resistance, corporate management in the United States began replacing the executives in these foreign divisions with American executives who could change the corporate culture to one that would accept project management. This process took almost three years to complete.

Multinational cultures have serious issues that go well beyond executive recognition of the need to change and even the desire to do so. The following are typical multinational cultural complexities that must be addressed when using the PMMM. For simplicity's sake, they are listed according to the PMBOK® Guide (Project Management Book of Knowledge) processes.

Integration Management:
- Poor understanding of the benefits of project management
- Poor understanding of the role of the project manager
- No project management methodology
- Limited cross-functional access to certain groups
- Difficulty in gaining agreement and commitment
- Poor problem-solving capability

Scope Management:
- Improper assumptions
- Methodology designed for a single, national project
- Culture impacted by country's legal system
- No definition of a charter, scope baseline, or scope statement
- Education and experience of planners can vary

Time Management:
- Estimating using an 8-hour day may be unrealistic
- Different interpretation about assumptions
- Inaccurate or obsolete scheduling tools
- Time may not be viewed as a critical constraint
- Missed milestones are an acceptable practice
- Nonexistent templates

- Size of resource pool can vary
- Calendars are different concerning holidays and vacations

Cost Management:

- Estimating can be impacted by currency conversion and inflation rates
- Politics can impact the award of subcontracts
- Government instability can cause changes in or cancellation of projects

Procurement Management:

- Negotiation takes place per local customers
- Infrastructure causes delays in procurement resulting in damaged goods
- Specifications are improperly interpreted
- Gratuities and kickbacks may be common practice
- Use of resources may need government permission
- Source selection may be based upon culture and politics rather than requirements and quality of suppliers

Risk Management:

- Ineptness and withholding of information
- Virtually no tools for risk management
- Too much government intervention
- Inaccurate understanding of problems
- Identifying risks may end one's career

Quality Management:

- Differing codes and laws
- Inadequate skill levels
- Differing view of quality
- No quality policies

Human Resource Management:

- Inadequate skill levels and training
- Political instability in human resource practices
- Decisions based upon holidays, customs, and dietary considerations
- Differing value systems

Communications Management:

- Increased barriers and filters
- Poor understanding of each party's language
- Time zone differences
- Lack of trust
- Poorly understood speech etiquette
- Misinterpretation of body language

PARTNERSHIPS

A partnership is a group of two or more individuals or groups working together to achieve a common objective. Historically, partnerships were often between companies working together rather than between individuals. In project management, the critical partnership is the working relationship between the project and line managers.

In the early days of project management, the selection of the individual to serve as the project manager was most often dependent upon who possessed the greatest command of technology. The ultimate result, as shown in Figure 11–10, was a very poor working relationship between the project and line managers. Line managers viewed project managers as a threat, and their relationship developed into a competitive, superior-subordinate relationship. The most common form of organizational structure was a strong matrix where the project manager, perceived as having a command of technology, had a greater influence over the assigned employees than did their line manager.

As the magnitude and technical complexity of the projects grew, it became obvious that the project managers could not maintain a command of technology in all aspects of the project. Now, project managers were viewed as possessing an understanding of rather than command of technology. Project managers were now becoming more dependent upon line managers for technical support. The project manager now finds himself or herself in the midst of a weak matrix where the employees are receiving the majority of their technical direction from the line managers.

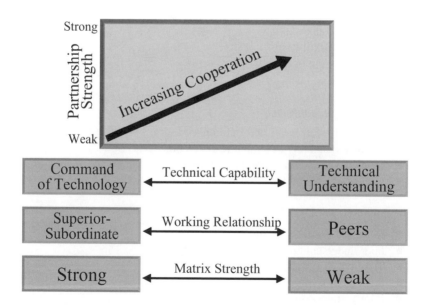

FIGURE 11–10. Partnership strength.

As the partnership between the project and line managers begins to develop, management recognizes that partnerships work best on a peer-to-peer basis, rather than on a superior-to-subordinate basis. Now the project and line managers view each other as equals and share in the authority, responsibility, and accountability needed to assure project success.

Good project management methodologies emphasize the cooperative working relationship that must exist between the project and line managers. Conflict resolution channels may be clearly spelled out as part of the methodology, and which manager has the final decision in specific areas of knowledge may be specified.

THE IMPACT OF RISK CONTROL MEASURES

Most project management methodologies today have sections on risk management. The project management methodology may very well dictate the magnitude of the risk control measures to be undertaken. The risk control measures for risk assumption may be significantly more complex than risk control measures for avoidance.

Figure 11–11 shows the intensity of the controls versus the intensity of the risks. As the intensity of the risks increase, we tend to place more controls in the risk management process and the project management methodology. Care must be taken that the cost of maintaining these control measures does not overly burden the project. Time and money are required for effective risk management.

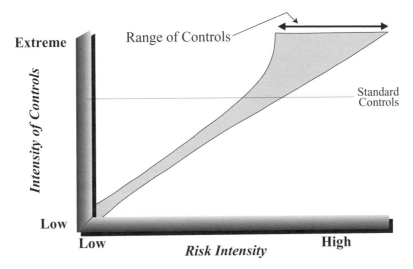

FIGURE 11–11. Risk control measures.

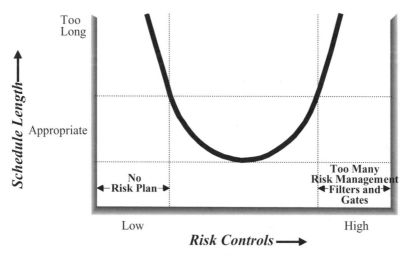

FIGURE 11–12.   Risk controls.

Excessive controls may require that the project manager spend more time performing risk management rather than actually managing the project.

How to determine the proper amount of risk control measures is not easy. This can be seen from Figure 11–12, which illustrates the impact on the schedule constraint. If not enough control measures are in place, or if there simply is no risk management plan, the result may be an elongated schedule due to ineffective risk control measures. If excessive risk control measures are in place, such as too many filters and gates, the schedule can likewise be elongated because the workers are spending too much time on contingency planning, risk reporting, documentation, and risk management meetings (i.e., there may be too many gates). This results in very slow progress being made. A proper balance is needed.

DEPENDENCIES BETWEEN RISKS

If project managers had unlimited funding, they could always identify a multitude of risk events. Some of the risk impacts may be insignificant, whereas others may expose the project to severe danger. With a large number of possible risk events, it is impossible to address each and every situation. It may be necessary to prioritize risks.

Assume that the project manager categorizes the risks according to the project's time, cost, and performance constraints, as illustrated in Figure 11–13. According to the figure, the project manager should focus his/her efforts on reducing the scheduling risks first. The prioritization of risks could be established by the project manager, by the project sponsor, or even by the customer. The pri-

	Schedule	Cost	Technical Performance or Quality
First (Highest) Priority	✓		
Second Priority			✓
Third Priority		✓	

FIGURE 11–13. Prioritization of risks.

oritization of risks can also be industry- or even country-specific, as shown in Figure 11–14. It is highly unlikely that any project management methodology would dictate the prioritization of risks. It is simply impossible to develop standardization in this area such that the application could be uniformly applied to each and every project.

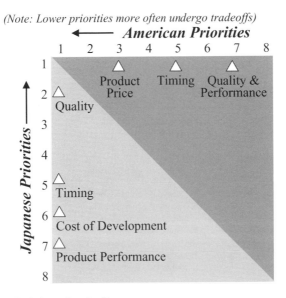

FIGURE 11–14. Ordering of tradeoffs.

The prioritization of risks for an individual project is a good starting point and could work well if it were not for the fact that most risks are interrelated. We know from tradeoff analysis that changes to a schedule can, and probably will, induce changes in cost and performance. Therefore, even though schedules have the highest priority in Figure 11–13, risk response to the schedule risk events may cause immediate evaluation of the technical performance risk events. Risks are interrelated.

The interdependencies between risks can also be seen from Table 11–6. The first column identifies certain actions that the project manager can opt for to take advantage of the opportunities in column 2. Each of these opportunities, in turn, can cause additional risks, as shown in column 3. In other words, risk mitigation strategies that are designed to take advantage of an opportunity could create another risk event that is more severe. As an example, working overtime could save you $15,000 by compressing the schedule. But if the employees make more mistakes on overtime, retesting may be required, additional materials may need to be purchased, and a schedule slippage could well occur, thus causing a loss of $100,000. Therefore, is it worth risking a loss of $100,000 to save $15,000?

To answer this question, we can use the concept of expected value, assuming we can determine the probabilities associated with mistakes being made and the cost of the mistakes. Without any knowledge of these probabilities, the actions taken to achieve the opportunities would be dependent upon the project manager's tolerance for risk.

Most project management professionals seem to agree that the most serious risks, and the ones about which we seem to know the least, are the technical risks. The worst situation is to have multiple technical risks that interact in an unpredictable or unknown manner.

As an example, assume you are managing a new product development project. Marketing has provided you with two technical characteristics that would make the product highly desirable in the marketplace. The exact relationship between these two characteristics is unknown. However, your technical subject matter experts

TABLE 11–6.   RISK INTERDEPENDENCIES

Action	Opportunity	Risk
• Work overtime	• Schedule compression	• More mistakes; higher cost and longer schedule
• Add resources	• Schedule compression	• Higher cost and learning curve shift
• Arrange for parallel work	• Schedule compression	• Rework and higher costs
• Reduce scope	• Schedule compression and lower cost	• Unhappy customer and no follow-on work
• Hire low cost resources	• Lower cost	• More mistakes and longer time period
• Outsource critical work	• Lower cost and schedule compression	• Contractor possesses critical knowledge at your expense

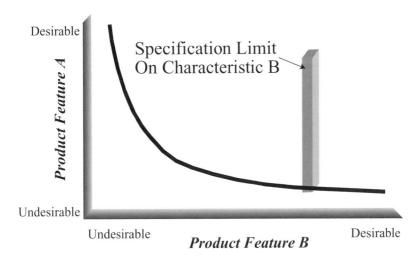

FIGURE 11–15. Interacting risks.

have prepared the curve shown in Figure 11–15. According to the curve, the two characteristics may end up moving in opposite directions. In other words, maximizing one characteristic may require degradation in the second characteristic.

Working with marketing, you prepare the specification limits according to characteristic B in Figure 11–15. Because these characteristics interact in often unknown ways, the specification limit on characteristic B may force characteristic A into a region that would make the product less desirable to the ultimate consumer.

Although project management methodologies provide a framework for risk management and the development of a risk management plan, it is highly unlikely that any methodology would be sophisticated enough to account for the identification of technical dependency risks. The time and cost associated with the identification, quantification, and handling of technical risk dependencies could severely tax the project financially.

Another critical interdependency is the relationship between change management and risk management, both of which are part of the singular project management methodology. Each risk management strategy can results in changes that generate additional risks. Risks and changes go hand in hand, which is one of the reasons companies usually integrate risk management and change management together into a singular methodology. Table 11–7 shows the relationship between managed and unmanaged changes. If changes are unmanaged, then more time and money is needed to perform risk management. And what makes the situation even worse is that higher salaried employees and additional time is required to assess the additional risks resulting from unmanaged changes. Managed changes, on the other hand, allows for a lower cost risk management plan to be developed.

TABLE 11–7.   UNMANAGED VERSUS MANAGED CHANGES

Type of Change	Where Time Is Invested	How Energy Is Invested	Which Resources Are Used
Unmanaged	• Back-end	• Rework • Enforcement • Compliance • Supervision	• Senior management and key players only
Managed	• Front-end	• Education • Communication • Planning • Improvements • Value added	• Stakeholders (internal) • Suppliers • Customers

Project management methodologies, no matter how good, cannot accurately define the dependencies between risks. It is usually the responsibility of the project team to make these determinations.

SELECTING THE APPROPRIATE RESPONSE MECHANISM

There exist four widely accepted risk response methods; assumption, reduction, transfer, and avoidance. Historically, most practitioners argue that the risk response method selected is heavily biased toward the magnitude of the risk and the project manager's tolerance level for risk. While this may still be true, there are other factors that influence the risk response method selected, and many of these can be included as part of the project management methodology.

The potential rewards of selecting the appropriate risk response can influence the selection process. Figure 11–6 shows the risk-reward matrix. What is important to recognize in Figure 11–16 is that the risk-reward matrix is actually three-dimensional, with the third axis being the quality of resources required. Certain risk response actions, such as assumption or reduction, require that resources be consumed. The quality and availability of the resources required can influence the risk response selection process irrespective of the potential rewards. For example, if a company adopts a risk assumption approach on an R&D project, the rewards could be huge if patents are issued and licensing agreements follow. But this is predicated upon available, qualified resources. Without the appropriate available resources, the only response mechanisms remaining might be risk avoidance or risk transfer.

A second factor influencing the risk response method selected is the procedural documentation requirements of the project management methodology. This appears in Figure 11–17. Project management methodologies that are based on

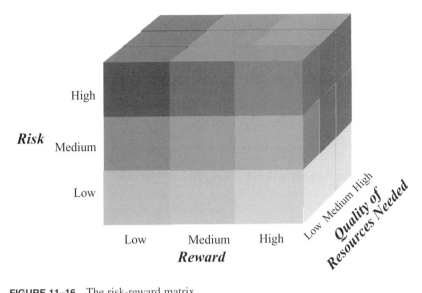

FIGURE 11–16. The risk-reward matrix.

policies and procedures are very rigid. Most good methodologies today are based upon guidelines that provide the project manager with significantly more flexibility in decision-making.

This flexibility (i.e., use of guidelines) can affect the risk response method selected. Although no empirical data exists as yet to support this, there appears a

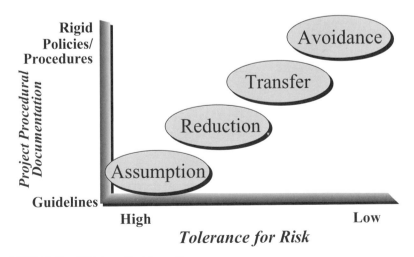

FIGURE 11–17. Which method to use?

tendency for project managers to accept higher levels of risk if the project manager is given more freedom in decision-making. On the other hand, the rigidity of policies and procedures generally allows for lower levels of risk acceptance and project managers seem to prefer avoidance. As risk management grows, more research can be expected in this area.

CONCLUSIONS

General strategic planning is never accomplished only once. It must be reexamined over and over again as requirements change and new information appears. The same holds true for strategic planning for project management. As more and more of these problem areas are identified internally from lessons learned files and externally in published articles, some of these often overlooked problems will be corrected.

The Project Office*

INTRODUCTION

As companies begin to recognize the favorable effect that project management has on profitability as well as on customer satisfaction, emphasis is placed upon strategic planning for project management using the project office concept. The concept of a project office (PO) or project management office (PMO) could very well be the most important project management activity in this decade. With this recognition of importance comes strategic planning for both project management and the project office. Maturity and excellence in project management does *not* occur simply by using project management over a prolonged period of time. Rather, it comes through strategic planning for both project management and the project office.

General strategic planning involves the determination of where you wish to be in the future and then how you plan to get there. For project office strategic planning, it is often easier to decide which activities should be under the control of the project office than to determine how or when to make the necessary changes. For each activity placed under the auspices of the project office, there may appear pockets of resistance that initially view removing this activity from their functional area as a threat to their power and authority.

---

* This chapter has been adapted from Harold Kerzner, *Advanced Project Management: Best Practices in Implementation.* New York: Wiley, 2004, Chapter 8. Reproduced by permission of John Wiley & Sons.

THE PROJECT OFFICE: 1950–1990 _____

For almost 40 years, the project office (or program office) functioned as a customer group project office and was compromised of a group of project management personnel assigned to a specific project, usually a large project. Aerospace and defense contractors often created three separate project offices, each serving Army, Navy, or Air Force customers. Some companies created project offices exclusively to service either large projects or small projects. Each project office performed its own strategic planning, resulting in suboptimization and a lack of synergy.

The concept behind this project office approach was to get closer to the customer by setting up an organization dedicated to that customer. Strategic planning focused heavily on customer relations and future business with that customer. The project office became an organization within an organization and could function as a "real" or "virtual" organization to service a particular customer. The majority of these so-called projects were actually programs that were very large in dollar value and that had multiyear government funding. It was not uncommon for people to spend 10 or 15 years working on just one project.

The members of the project office had unique roles and responsibilities, but essentially worked together as a project management team. Each person in the project office may have been required to have both primary and secondary responsibilities. The secondary responsibilities included functioning as a backup for other project office personnel in case some project office personnel were reassigned to other projects, left the company, or were out sick.

Headcount in the project office was not considered extremely important because the customer often paid any added costs. Technology and schedules were viewed as significantly more important than cost. Customers preferred to have more people than necessary assigned to project offices. The cost of having someone assigned full-time to the project office was viewed as an insignificant overmanagement cost compared to the risks of undermanagement, where individuals were assigned part-time but might be needed full-time. The only people who were trained in project management, and truly understood it, were the project office personnel. Project offices functioned horizontally throughout the organization and were viewed as profit centers, whereas the traditional functional hierarchy was treated as a cost center. Strategic planning emphasized near-term profitability.

During the 1980s, military and government agencies became more cost-conscious. Project offices were pared down as personnel other than those assigned to the project office underwent training in project management. Line managers also underwent training and were asked to better understand project management and share accountability with project managers for project success.

THE PROJECT OFFICE: 1990–2000 _____

The 1990s began with a recession that took a heavy toll on white-collar ranks. Management's desire for efficiency and effectiveness led them to take a hard look at nontraditional management techniques such as project management. Project management began to expand to non–project-driven industries. The benefits of using project management, which were once seen as applicable only to the aerospace, defense, and heavy construction industries, were now recognized as being applicable for other industries. The importance of allowing the project office to perform strategic planning for project management was becoming apparent.

As the benefits of project management became apparent, management understood that there might be a significant, favorable impact on the corporate bottom line. This led management to two important conclusions:

- Project management had to be integrated and compatible with the corporate reward systems for sustained project management growth
- Corporate recognition of project management as a profession was essential in order to maximize performance

The recognition of project management professionalism led companies to accept the Project Management Institute's (PMI's) Certification Program for project management personnel as the standard and to recognize the importance of the project office concept. Consideration was given to placing all critical activities related to project management under the supervision of the project office. This included such topics as:

- Standardization in estimating
- Standardization in planning
- Standardization in scheduling
- Standardization in control
- Standardization in reporting
- Clarification of project management roles and responsibilities
- Preparation of job descriptions for project managers
- Preparation of archive data on lessons learned
- Continuous benchmarking
- Development of project management templates
- Development of a project management methodology
- Recommending and implementing changes and improvements to the existing methodology
- Identification of project standards
- Identification of best practices
- Performing strategic planning for project management
- Establishment of a project management problem-solving hotline
- Coordinating and/or conducting project management training programs
- Transfer of knowledge through coaching and mentorship
- Development of a corporate resource capacity/utilization plan
- Risk assessment
- Disaster recovery planning

With these changes taking place, organizations began changing the name of the project office to the Center of Excellence (COE) in project management. The COE was mainly responsible for providing information to stakeholders rather than for actually executing projects or making midcourse corrections to a plan.

Each of the activities in the above list brought with it both advantages and disadvantages. The majority of the disadvantages could be attributed to the increased levels of resistance to the new responsibilities given to the project office.

For simplicity's sake, the resistance levels can be classified as low risk, moderate risk, and high risk, according to the following definitions:

- *Low risk:* Easily accepted by the organization with very little shift in the balance of power. Virtually no impact on the corporate culture.
- *Moderate risk:* Some resistance by the corporate culture and possibly a shift in the balance of power and authority. Resistance levels can be overcome in the near term and with minimal effort.
- *High risk:* Heavy pockets of resistance exist and a definite shift in some power and authority relationships. Strong executive leadership may be necessary to overcome the resistance.

Not every project office has the same responsibilities. Likewise, the same responsibilities implemented in two project offices can have differing degrees of risk. People tend to resist change even when they know the change may be in the best interest of the organization.

Figure 12–1 shows typical risk levels for implementing the selected project office responsibilities. Evaluating potential implementation risks is critical. It may be easier to gain support for the establishment of a project office by implementing low risk activities first. The low risk activities in Figure 12–1 are operational activities to support project management efforts in the near term, whereas the high risk activities are more in line with strategic planning responsibilities and possibly the control of sensitive information.

THE PROJECT OFFICE: 2000–PRESENT

As we entered the twenty-first century, the project office became commonplace in the corporate hierarchy. Although the majority of activities assigned to the project office did not change, there was now a new mission for the project office:

> *The project office now has the responsibilities of maintaining all intellectual property related to project management and to actively supporting corporate strategic planning.*

Low	Moderate	High
•Mentorship	•Strategic planning	•Capacity planning
•Standards	•Lessons learned files	•Benchmarking
•Training	•Continuous	•Dissemination of
•Stakeholder	improvement	information
management	•Performance failure	•Business case
	reporting	development

FIGURE 12–1. Activity implementation risks.

The project office was now servicing the corporation, especially its strategic planning activities, rather than focusing on a specific customer. The project office was transformed into a corporate center for control of project management intellectual property. This was a necessity as the magnitude of project management information grew almost exponentially throughout the organization. Senior managers were now recognizing that project management and the project office had become an invaluable asset for senior management as well as for the working levels.

During the past ten years, the benefits for the executive levels of management of using a project office have become apparent. They include:

- Standardization of operations
- Company rather than silo (functional) decision-making
- Better capacity planning (i.e., resource allocations)
- Quicker access to higher quality information
- Elimination or reduction of company silos
- More efficient and effective operations
- Less need for restructuring
- Fewer meetings that rob executives of valuable time
- More realistic prioritization of work
- Development of future general managers

All of the above benefits are either directly or indirectly related to project management intellectual property. To maintain the project management intellectual property, the project office must maintain the vehicles for capturing the data and then disseminating the data to the various stakeholders. These vehicles include the company project management intranet, project web sites, project databases, and project management information systems. Since much of this information is necessary for both project management and corporate strategic planning, there must exist strategic planning for the project office.

TYPES OF PROJECT OFFICES

There exist three types of project offices commonly used in companies.

- **Functional project office:** This type of project office is utilized in one functional area or division of an organization, such as information systems. The major responsibility of this type of project office is to manage a critical resource pool, that is, to provide resource management.
- **Customer group project office:** This type of project office provides better customer management and customer communications. Common customers or projects are clustered together for better management and customer relations. More than one customer group project office can exist at the same time, and any of them may end up functioning as a temporary organization. In effect, this acts like a company within a company.
- **Corporate project office:** This type of project office services the entire company and focuses on corporate and strategic issues rather than functional issues.

More than one type of PMO can exist concurrently within a company. Some companies have functional PMOs just for the information technology organization as well as having a corporate PMO for strategic issues. The PMOs work together, especially during strategic planning efforts.

Multinational companies have regional PMOs dispersed throughout the world. These PMOs are all networked together through one centralized PMO that is responsible for the coordination of all regional PMOs.

PROJECT MANAGEMENT INFORMATION SYSTEMS

Given the fact that the project office is now the guardian of project management intellectual property, especially the knowledge needed for strategic planning activities, there must exist processes and tools for capturing this information. This information can be collected through four information systems, as shown in

FIGURE 12–2. Project management information systems.

Figure 12–2. Each information system can be updated and managed through the company intranet.

Earned Value Measurement Information System

The earned value measurement information system is common to almost all project managers. It provides sufficient information to answer two questions:

- Where is the project today?
- Where will the project end up?

This system either captures or calculates the planned and actual value of the work, the actual costs, cost and schedule variances (in hours or dollars, and percent), the estimated cost at completion, the estimated time at completion, percent complete, and trends.

The Earned Value Measurement Information System is critical for a company that requires readily available information for rapid decision-making. It is easier to make small changes to a project plan, rather than large ones. Therefore, variances from the performance management baseline must be identified quickly so that corrective action can be taken in small increments.

Risk Management Information System

The second information system provides data on risk management. The risk management information system (RMIS) stores and allows retrieval of risk-related data. It provides data for creating reports and serves as the repository for all current and historical information related to project risk. The information will include risk identification documentation (possibly created by using templates), quantitative and qualitative risk assessment documents, contract deliverables if appropriate, and any other risk-related reports. The project management office will use the data from RMIS to create reports for senior management and retrieve data for the day-to-day management of projects. By using risk management templates, each project will produce a set of standard reports for periodic reporting and have the ability to create ad hoc reports in response to special queries. This information is directly related to the failure reporting information system and the lessons learned information system.

The Performance Failure Information System (PFIS)

The project office may have the responsibility for maintaining the performance failure information system. The failure could be a complete project failure or simply the failure of certain tests within the project. The PFIS must identify the cause(s) of the failure and possibly recommendations for the removal of the cause(s). The cause(s) could be identified as coming from problems entirely internal to the organization or from coordinated interactions with subcontractors.

It is the project office's responsibility to develop standards for maintaining the PFIS rather than for validating the failure. Validation is the responsibility of the team members performing the work. Failure reporting can lead to the discovery of additional and more serious problems. First, there may be resistance to report some failures for fear that they may reflect badly upon the personnel associated with the failure, such as the project sponsors. Second, each division of a large company may have its own procedures for recording failures and may be reluctant to make the failure visible in a corporate-wide database. Third, there could exist many different definitions of what is or is not a failure. Fourth, the project office may be at the mercy of others to provide accurate, timely, and complete information.

The failure report must identify the item that failed, symptoms, conditions at the time of the failure, and any other pertinent evidence necessary for corrective action to take place. Failure analysis, which is the systematic analysis of the consequences of the failure on the project, cannot be completed until the cause(s) of the failure has (have) been completely identified. The project office may simply function as the records keeper to standardize a single company-wide format and database for reporting the results of each project. This could be part of the lessons learned review at the end of each project.

Consider the following example: An aerospace company had two divisions that often competed with one another through competitive bidding on government contracts. Each conducted its own R&D activities and very rarely exchanged data. One of the divisions spent six months working on an R&D project that was finally terminated and labeled as a failure. Shortly thereafter, it was learned that the sister division had worked on exactly the same project a year before and achieved the same unproductive results. Failure information had not been exchanged, resulting in the waste of critical resources.

Everyone recognizes the necessity for a corporate-wide information system for storing failure data. But there always exists the risk that some will view this as a loss of power. Others may resist for fear that their name will be identified along with the failure. The overall risk with giving this responsibility to the project office is low to moderate.

Lessons Learned (Postmortem Analysis) Information System

Some companies work on a vast number of projects each year, and each of these projects provides valuable information for improving standards, estimating for future bidding, and improving the way business is conducted. All of this information is intellectual property and must be captured for future use. Lesson learned reviews are one way to obtain this information.

If intellectual property from projects is to be retained in a centralized location, then the project office must develop expertise in how to conduct a postmortem analysis meeting. At that meeting, four critical questions must be addressed:

- What did we do right?
- What did we do wrong?
- What future recommendations can be made?
- How, when, and to whom should the information be disseminated?

Additional questions that must be asked follow the postmortem pyramid shown in Figure 12–3. The objectives for a project are established from the top of the pyramid to the bottom. However, the postmortem analysis that evaluates the project's metrics or measurements goes in reverse order, from the bottom to the top. The bottom level, which is the basic level, evaluates the deliverables in terms of time, cost, quality, and scope. These constraints are often referred to as the critical success factors (CSFs), and are considered as seen through the eyes of the client.

Typical questions to consider for the critical success factors include:

Time

- Were the schedules realistic?
- Was the level of detail correct?
- Was it easy to evaluate performance from the schedule?
- Was tracking easily accomplished?

Cost

- How accurate were our estimates?
- Do our estimates need to be updated?
- Did cost tracking follow our methodology?
- Were there any problems with cost reporting?

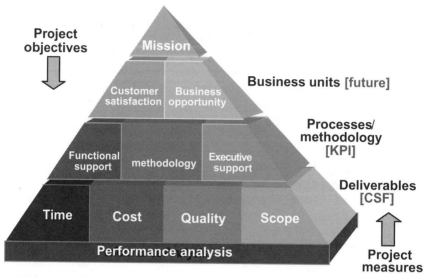

FIGURE 12-3. The postmortem pyramid

Quality

- Did we conform to the customer's specifications?
- Did the product perform as expected?
- Did we evaluate durability, reliability, serviceability and aesthetics?

Scope

- Was the statement of work (SOW) easily understood?
- Were the objectives clearly defined?
- Was there proprietary technology involved?
- If so, does the company have patent protection?
- How difficult were the tradeoffs?

The second layer in the postmortem pyramid of Figure 12–3 contains the key performance indicators (KPIs). KPIs are the internally shared learning topics that allow us to maximize what we do right and correct what we do wrong. KPIs are the "internal best practices" that allow us to achieve the critical success factors. Success is normally defined in terms of both CSFs and KPIs.

The KPIs can be categorized into the three areas shown in Figure 12–3. Typical questions for each KPI area might include:

Line Management (Functional) Support

- Did the assigned personnel have the required expertise?
- What was the quality of the assigned resources?

- Did the resources possess innovative capability?
- Was the right quantity of resources assigned?
- Were the resources assigned in a timely manner to support the schedule?
- Was there resource overload?

Methodology

- Did the methodology allow for quick response?
- Was the planning performed correctly?
- Did the methodology allow for contingency planning?
- Were the tools to support the methodology available and state-of-the-art?

Senior Management (Executive) Support

- Did senior management function as a sponsor?
- Were they helpful?
- Did they decentralize decision-making?
- Did the project team have sufficient authority for the work required?
- Was there a charter?

The third layer in the postmortem pyramid of Figure 12–3 is the business unit evaluation. This evaluation focuses on two areas: customer satisfaction and future business opportunities. Typical questions for these areas include:

Customer Satisfaction

- Was the customer pleased with the price–quality–value relationship?
- Were the deliverables provided on time?
- Are there value-added opportunities or follow-on work available?

Business Opportunities

- Were our preconceptions valid?
- Are there additional sales opportunities other than with this client?
- Will this project allow the organization to grow?
- Will this project fulfill the mission?

DISSEMINATION OF INFORMATION

A problem facing most organizations is how to make sure that critical information, such KPIs and CSFs, are known throughout the organization. Intranet lessons learned databases would be one way to share information. However, a better way might be for the project office to take the lead in preparing lessons learned case studies at the end of each project. The case studies could then be used in future training programs throughout the organization and be intranet-based.

As an example, a company completed a project quite successfully, and the project team debriefed senior management at the end of the project. The company had made significant breakthroughs in various manufacturing processes used for the project, and senior management wanted to make sure that this new knowledge would be available to all other divisions.

The decision was made to dissolve the team and reassign its members to various divisions throughout the organization so that they could disseminate what they had learned working on the project. After six months had passed, however, it became evident that very little of that knowledge had been transmitted to the other divisions. The team was then reassembled and asked to write a lessons learned case study to be used during project management training programs.

Although this approach worked well in this instance, there also exist detrimental consequences that may make this approach difficult to implement. For example, another company had adopted the concept of preparing lessons learned case studies. Although the end result of one of the projects was a success, several costly mistakes were made during the execution of the project, due to a lack of knowledge of risk management and poor decision-making. Believing that lessons learned case studies should include mistakes as well as successes, the project office team preparing the case study included all information.

Despite attempts to disguise the names of the workers who had made the critical mistakes, everyone in the organization knew who had worked on the project and was able to discover who the employees were. Several of the workers involved in the project filed a grievance with senior management over the disclosure of this information, and the case studies were then removed from training programs. It takes a strong organizational culture to learn from mistakes without retribution to the employees. Thus having the project office administer this activity may result in moderate to high risk.

MENTORING

Project management mentoring is a critical project office activity. Most people seem to agree that the best way to train someone in project management is with on-the-job training. One way to implement such training is to have inexperienced project managers work directly under the guidance of an experienced project manager, especially on large projects. This approach may become costly if the organization does not have a stream of large projects.

Perhaps the better choice would be for the project office to assume a mentoring role whereby inexperienced project managers can seek advice and guidance from the more experienced project managers who report either "solid" or "dotted" to the project office. This approach has three major benefits. First, the line manager to whom the project manager reports administratively may not have the necessary project management knowledge or experience, and thus may not be capable of assisting the worker in times of trouble. A more experienced project manager, acting as mentor, can fill this gap. Second, the project manager may not wish to discuss certain problems with his or her superior for fear of retribution, but may be quite willing to seek help elsewhere. Third, given the fact that the project office may have the responsibility for maintaining lessons learned files, the project mentoring program could use these files and provide the inexperienced project manager with early warning indicators that potential problems could occur.

The mentoring program could be done on a full-time basis or on an as-needed basis. The latter is usually the preferred approach. Full-time mentoring may seem like a good idea, but it includes the risk that the mentor will end up managing the project. The overall risk for project office mentoring is low.

DEVELOPMENT OF STANDARDS AND TEMPLATES

A critical component of any project office is the development of project management standards. Standards foster teamwork by creating a common language. However, developing excessive standards in the form of policies and procedures may be a mistake because it may not be possible to create policies and procedures that cover every possible situation on every possible project. In addition, the time, money, and people required to develop rigid policy and procedure standards would make project office implementation unlikely because of headcount requirements.

Forms and checklists can be prepared in a template format such that the information can be used on a multitude of projects. Templates should be custom-designed for a specific organization rather than copied from another organization that may not have similar types of projects or a similar culture. Reusable templates should be prepared *after* the organization has completed several projects, whether successfully or unsuccessfully. At that point lessons learned information can be used for the development and enhancement of the templates.

There is a danger, however, in providing templates as a replacement for the more formalized standards. First of all, because templates serve as a guide for a general audience, they may not satisfy the needs of any particular program. Second, there is the risk that some prospective users of the templates, especially inexperienced project managers, may simply adopt the templates "as required as written" despite the fact that they do not fit a particular program.

The reason for providing templates is *not* to tell the team how to do their job, but to give the project manager and his/her staff a starting point for their own project initiation, planning, execution, control, and closure processes. Templates should stimulate proactive thinking about what has to be done and possibly some ideas on how to do it. Templates and standards often contain significantly more information than most project managers need. However, the templates and standards should be viewed as the key to keeping things simple, and the project managers should be able to tailor the templates and standards to suit the needs of the project by focusing on the key critical areas.

Templates and standards should be updated as necessary. Since the project office is most likely responsible for maintaining lessons learned files and project postmortem analysis, it is only fitting that the project office evaluate these data to seek out KPIs that could dictate template enhancements. Standards and templates can be regarded as a low-risk project office activity.

Templates and standards are very important to consulting companies, not necessarily for internal use, but rather for the benefit of their clients. The standards and templates must be flexible enough to be adapted easily to the needs of different clients.

PROJECT MANAGEMENT BENCHMARKING

Perhaps the most interesting and most difficult activity assigned to a project office is benchmarking. Just like mentoring, benchmarking requires the use of experienced project managers. The assigned individuals must know what to look for and what questions to ask, must have the ability to recognize a good fit with the company, must understand how to evaluate the data, and must be able to determine what recommendations to make.

Benchmarking is directly related to strategic planning for project management, and it can have a pronounced effect on the corporate bottom line, depending on how quickly the recommended changes are implemented. In recent years, companies have discovered that best practices can be benchmarked against organizations not necessarily in the same line of business. For example, an aerospace division of a large firm had been using project management for over 30 years. During the early 1990s, the firm had been performing benchmarking studies but *only* against other aerospace firms. Complacency had set in, with the firm believing that they were on an equal standing with their competitors in the aerospace field. In the late 1990s, the firm began benchmarking against firms outside of their industry, specifically firms in the telecommunications, computers, electronics, and entertainment fields. Most of these firms had been using project management for less than five years and, in that time, had achieved project management performance that exceeded that of the aerospace firm. Now the aerospace firm benchmarks against all industries.

Project office networking for benchmarking purposes could very well be in the near future for most firms. Project office networking could span industries and continents. In addition, it may become commonplace even for competitors to share project management knowledge. However, at present, it appears that the majority of project management benchmarking is being performed by organizations whose function is entirely benchmarking. These organizations charge a fee for their services and conduct symposiums for their membership whereby project management best practices data are shared. In addition, they offer database services through which you can compare your organization against:

- Other organizations in your industry
- Other organizations in different industry sectors

- Other employee responses within your company
- Other organizations by company size
- Other organizations by project size

Some organizations have a strong resistance to benchmarking. The arguments against benchmarking include:

- It doesn't apply to our company or industry.
- It wasn't invented here.
- We're doing fine! We don't need it.
- Let's leave well enough alone.
- Why fix something that isn't broken?

Because of these concerns, benchmarking may be a high-risk activity because of the fear of recommended changes.

BUSINESS CASE DEVELOPMENT

One of the best ways for a project office to support the corporate strategic planning function is by becoming experts in business case development. More specifically, the project office can develop expertise in feasibility studies and cost-benefit analysis. In the "Scope Management" section of the Project Management Book of Knowledge (PMBOK®) Guide, one of the outputs of the Scope Initiation Process is the identification/appointment of a project manager. This is accomplished after the business case is developed. There are valid arguments for assigning the project manager only after the business case is developed:

- The project manager may not be able to contribute to the business case development.
- The project might not be approved and/or funded, and it would be an added cost to have had the project manager on board early.
- The project might not be defined well enough to determine at an early stage the best person to be assigned as the project manager.

While these arguments seem to have merit, assigning the project manager after developing the business case can raise a more serious issue in that the person ultimately assigned may not have sufficient knowledge about the assumptions, constraints, and alternatives that were considered during the business case development, which could lead to a less than optimal project plan. It is wishful thinking to believe that the project charter, which may have been prepared by someone completely separated from the business case development efforts, contains all of the necessary assumptions, alternatives, and constraints.

One of the axioms of project management is that the earlier the project manager is assigned, the better the plan and the greater the commitment to the project. Companies argue that the project manager's contribution is limited during business case development. The reason for this belief is because the project managers have never been trained in how to perform feasibility studies and cost-benefit analysis. These courses are virtually nonexistent in the seminar marketplace.

Business case development often results in a highly optimistic approach, however, with little regard for the schedule and/or the budget. Pressure is then placed upon the project manager to accept arguments and assumptions made during business case development. If the project fails to meet business case expectations, then the blame is placed upon the project manager rather than on the unrealistic expectations.

The project office must develop expertise in feasibility studies, cost-benefit analysis, and business case development. This expertise lends itself quite readily to templates, forms, and checklists. The project office can then become a viable support arm to the sales force in helping them make more realistic promises to the customers, and may even assist in generating additional sales. In the future, the project office might very well become the company experts in feasibility studies and cost-benefit analyses, and eventually conduct customized training for the organization on these subjects. Marketing and sales personnel, who traditionally perform these activities, may view this as a high risk.

CUSTOMIZED TRAINING (RELATED TO PROJECT MANAGEMENT)

For years, the training branch of the human resources group had the responsibility of working with trainers and consultants in the design of customized project management training programs. While many of these programs were highly successful, many others were viewed as failures. One division of a large company recognized the need for training in project management. The training department went out for competitive bidding and selected a trainer. The training department then added in their own agendas after filtering all of the information concerning the goals and deliverables sought by the division requesting the training. The trainer never communicated directly with the organization requesting the training and simply designed the course around the information presented by the training department. The training program was viewed as a failure and the consultant/trainer was never invited back. Postmortem analysis reached the following conclusions:

- The training branch (and the requesting organization) never recognized the need to have the trainer meet directly with the requesting organization.
- The training group received input from senior management, unknown to the requesting organization, as to what information they wished to see covered, and the resulting course satisfied nobody's expectation.

- The trainer requested that certain additional information be covered while suggesting that other information was inappropriate and should be deleted. The request fell upon deaf ears.
- The training department informed the trainer that they wanted only lecture, with no case studies and minimal exercises. This was the way training was done in other courses this company had held. The participant evaluations complained about lack of exercises and case studies.

While the training group believed that their actions were in the company's best interest, the results were devastating. The trainer was also at fault for allowing this situation to exist.

Successful project management implementation has a positive effect on corporate profitability. Given that this is true, why allow nonexperts to design project management coursework? Even line managers who believe that their organization requires project management knowledge may not know what to stress and what not to stress from the PMBOK®.

The project office has the expertise in designing project management course content. The project office maintains intellectual property on lessons learned files and project postmortem analysis, giving the project office valuable insight on how to obtain the best return on investment on training dollars. This intellectual property could also be invaluable in assisting line managers in designing courses specific to their organization. This activity is a low risk for the project office.

MANAGING STAKEHOLDERS

All companies have stakeholders. Figure 12–4 depicts the broad range of stakeholders, which for simplicity's sake, have been categorized as organizational, product/market, and capital markets. Apprehension may exist in the minds of some individuals that the project office will become the ultimate project sponsor responsible for all stakeholders. While this may happen in the future, it is highly unlikely that it will occur in the near term.

The project office focuses its attention on internal (organizational) stakeholders. It is not the intent of a project office to replace executive sponsorship. As project management matures within an organization, it is possible that not all projects will require executive sponsorship. In such situations, the project office (and perhaps middle management) may be given the added responsibility of some sponsorship activities, but most likely for internal projects.

The project office is a good "starting point" for building and maintaining alliances with key stakeholders. However, the project office's activities are designed to benefit the entire company, and giving the project office sponsorship responsibility may create a conflict of interest for project office personnel. Partnerships with key stakeholders must be built and nurtured, and that requires time. Stakeholder management may rob the project office personnel of valuable time needed for other activities. The overall risk for this activity is low.

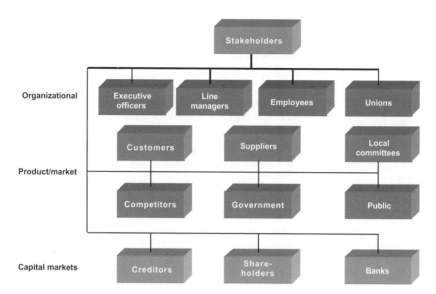

FIGURE 12–4. Project stakeholders.

CONTINUOUS IMPROVEMENT

Given the fact that the project office is a repository of project management intellectual property, the project office may be in the best position to identify continuous improvement opportunities. The project office should not have unilateral authority for implementing the changes, but rather the ability to recommend changes. Some organizations maintain a strategic policy board or executive steering committee that, as one of its functions, evaluates project office continuous improvement opportunities.

CAPACITY PLANNING

Of all of the activities assigned to a project office, the most important activity in the eyes of senior management could very well be capacity planning. For executives to fulfill their responsibility as the architects of the corporate strategic plan, they must know how much additional work the organization can take on, when, and where, without excessively burdening the existing labor pool. The project office must work closely with senior management on all activities related to portfolio management and project selection. Strategic timing, which is the process of deciding which projects to work on and when, is a critical component of strategic planning.

Senior management could "surf" the company intranet on an as-needed basis to view the status of an individual project without requiring personal contact with the team. But to satisfy the requirements for strategic timing, all projects would need to be combined into a single database that identifies the following:

- Resources committed per time period and per functional area
- Total resource pool per functional area
- Available resources per time period per functional area

There may be some argument whether the control of this database should fall under the administration of the project office. The author believes that this should be a project office responsibility because:

- The data would be needed by the project office to support strategic planning efforts and project portfolio management.
- The data would be needed by the project office to determine realistic timing and costs to support competitive bidding efforts.
- The project office may be delegated the responsibility to determine resource skills required to undertake additional work.
- The data will be needed by the project office for upgrades and enhancements to this database and other impacted databases.
- The data may be necessary to perform feasibility studies and cost-benefit analysis.

This activity is a high-risk effort for the project office because line managers may see this as turf infringement.

RISKS OF USING A PROJECT OFFICE

Risks and rewards go hand-in-hand. The benefits of a project office can be negated if the risks of maintaining a project office are not effectively managed. Most risks do not appear during the creation of the project office, but more do so well after implementation. These risks include:

- **Headcount:** Once the organization begins recognizing the benefits of using a project office, there is a natural tendency to increase headcount in the project office with the false belief that additional benefits will be forthcoming. While this belief may be valid in some circumstances, the most common result is diminishing returns. As more of the organization becomes knowledgeable in project management, the headcount in the project office should decrease.
- **Burnout:** Employee burnout is always a risk. Using rotational or part-time assignments can minimize the risk. It is not uncommon for people

working in a project office to still report "solid" to their line manager and "dotted" to the project office,

- **Excessive paperwork:** Excessive paperwork costs millions of dollars to prepare and can waste precious time. Project activities work much better when using forms, guidelines, and checklists rather than the more rigid policies and procedures. To do this effectively, however, requires a culture based upon trust, teamwork, cooperation, and effective communications.
- **Organizational restructuring:** Information is power. Given the fact that the project office performs more work laterally than vertically, there can be power struggles for control of the project office, especially for control over the project managers. Project management and a project office can work quite well within *any* organizational structure that is based upon trust, teamwork, cooperation, and effective communications.
- **Trying to service everyone in the organization:** The company must establish some criteria for when the project office should be involved. Not all projects are monitored by the project office. The most common threshold limits for when to involve the project office include:
  - Dollar value of the project
  - Time duration of the project
  - Amount and complexity of cross-functionality
  - Risks to the company
  - Criticality of the project (i.e. cost reductions)

A critical question facing many executives is "How do we as executives measure the return-on-investment as a result of implementing a project office?" The actual measurement can be described in both qualitative and quantitative terms. Qualitatively, the executives can look at the number of conflicts coming up to the executive levels for resolution. With an effective project office acting as a filter, fewer conflicts should go up to the executive levels. Quantitatively, the executives can look at the following:

- **Progress reviews:** Without a project office, there may exist multiple scheduling formats, perhaps even a different format for each project. With a project office and standardization, the reviews are quicker and more meaningful.
- **Decision-making:** Without a project office, decisions are often delayed and emphasis is placed upon action items rather than meaningful decisions. With a project office, meaningful decisions are possible.
- **Wasted meetings:** Without a project office, executives can spend a great deal of time attending too many and very costly meetings. With a project office and more effective information, the executives can spend less time in meetings and more time dealing with strategic issues rather than operational issues.
- **Quantity of information:** Without a project office, the executives can end up with too little or too much project information. This may inhibit

effective decision-making. With a project office and standardization, executives find it easier to make timely decisions. The prime responsibility of senior management is strategic planning and deployment and worrying about the future of the organization. The prime responsibility of middle-level and lower-level management is to worry about operational issues. The responsibility of the project office is to act a bridge between all of the levels and make it easier for all levels to accomplish their goals and objectives.

REPORTING AND STRUCTURE

Not all companies support project management the same way. Despite the recognizable benefits that arise by using a project office, disagreements still exist in many organizations as to where the project office should report and how to get the most out of a project office. However, given the responsibilities of the project office and its relationship to corporate strategic planning, capacity planning, and project portfolio management, the project office must report to the executive levels of management. The shorter the distance between senior management and the project office, the more quickly the benefits of project management will be recognized.

Every company can have a different structure for its project office. A typical structure for a project office appears in Figure 12–5.

The intent of the project office is to manage intellectual property on project management and, as such, it should not create bureaucracy by adding layers of management. A project office does not need more than four or five individuals, assigned as seen in the Johnson Controls case study. In addition, individuals may be assigned part-time to a project office or could be "dotted" line reporting to the project office while still maintaining other functional responsibilities.

FIGURE 12–5. Simplified project office structure.

Because of the diversity of activities within a project office, individuals assigned to a project office could have multiple responsibilities and might serve as
a backup for one another. This would reduce the headcount in the project office
and might make it easier to measure the return-on-investment of using a project
office.

Six Sigma
and the Project
Management Office

IINTRODUCTION

In the previous chapter, we discussed the importance of the project management office (PMO) for strategic planning and continuous improvements. In some companies, the PMO has been established specifically for the supervision and management of Six Sigma projects. Six Sigma was originally developed as a means of reducing defects. Today, however, the meaning of Six Sigma has been expanded to include other activities as will be discussed in this chapter. Six Sigma teams throughout the organization gather data and make recommendations to the PMO for Six Sigma projects. The Six Sigma project manager, and possibly the team, would be permanently assigned to the PMO.

Unfortunately, not all companies have the luxury of maintaining a large PMO where the Six Sigma teams and other supporting personnel are permanently assigned to the PMO. It is the author's belief that the majority of the PMOs have no more than four of five people permanently assigned. Six Sigma teams, including the project manager, may end up reporting "dotted" to the PMO and administratively "solid" elsewhere in the organization. The PMO's responsibility within these organizations is primarily for the evaluation, acceptance, and prioritization of projects. The PMO may also be empowered to reject recommended solutions to Six Sigma projects.

For the remainder of this chapter we focus on organizations that maintain small PMO staffs. The people assigned to the PMO may possess a reasonable knowledge concerning Six Sigma but may be neither green nor black belts in Six Sigma. These PMOs can and do still manage selected Six Sigma projects, but perhaps not the traditional type of Six Sigma projects taught in the classroom.

TRADITIONAL VERSUS NONTRADITIONAL SIX SIGMA ⎯⎯⎯⎯

In the traditional view of Six Sigma, projects fall into two categories: manufacturing and transactional. Each category of Six Sigma is multifaceted and includes a management strategy, metric, and process improvement methodology. This is shown below in Figure 13–1. Manufacturing Six Sigma processes utilize machines to produce products, whereas transactional Six Sigma processes utilize people and/or computers to produce services. The process improvement methodology facet of Six Sigma addresses both categories. The only difference is what tools you will use. In manufacturing, where we utilize repetitive processes that make products, we are more likely to use advanced statistical tools. In transactional Six Sigma, we might focus more on graphical analysis and creative tools/techniques.

The traditional view of a Six Sigma project has a heavy focus on continuous improvement to a repetitive process or activity associated with manufacturing. This traditional view includes metrics, possibly advanced statistics, rigor, and a strong desire to reduce variability. Most of these Six Sigma projects fit better for implementation in manufacturing than in the PMO. These manufacturing-related projects are managed by Six Sigma teams.

Not all companies perform manufacturing and not all companies support the PMO concept. Companies without manufacturing needs might focus more on the transactional Six Sigma category. Companies without a PMO rely heavily upon the Six Sigma teams for the management of both categories of projects.

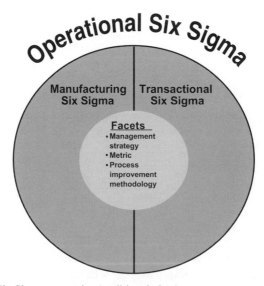

FIGURE 13–1. Six Sigma categories (traditional view).

Those companies that do support a PMO must ask themselves the following three questions:

- Should the PMO be involved in Six Sigma projects?
- If so, what type of project is appropriate for the PMO to manage even if the organization has manufacturing capability?
- Does the organization have sufficient resources assigned to the PMO to become actively involved in Six Sigma project management?

PMOs that are actively involved in most of the activities described in Chapter 12 do not have the time or resources required to support all Six Sigma projects. In such a case, the PMO must be selective as to which projects to support. The projects selected are commonly referred to as nontraditional projects, those that focus more on project management related activities than on manufacturing.

Figure 13–2 shows the nontraditional view of Six Sigma. In this view, operational Six Sigma includes manufacturing activities and all other activities from Figure 13–1, and transactional Six Sigma now contains primarily those activities to support project management.

In the nontraditional view, the PMO can still manage both traditional and nontraditional Six Sigma projects. However, there are some nontraditional Six Sigma projects that are more appropriate for management by the PMO. Some of the projects currently assigned to the PMOs include enhancements to the enterprise project management methodology, enhancements to the PMO tool set, efficiency improvements, and cost avoidance/reduction efforts. Another project assigned to

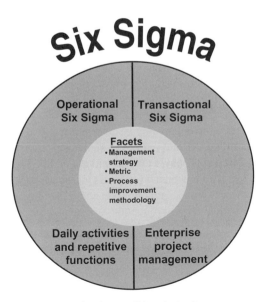

FIGURE 13–2. Six Sigma categories (nontraditional view).

the PMO involves process improvements to reduce the launch time of a new product and to improve customer management. Experts in Six Sigma might view these as nontraditional types of projects. There is also some concern as to whether these are really Six Sigma projects or just a renaming of a continuous improvement project to be managed by a PMO. Since several companies now refer to these as Six Sigma projects, the author will continue this usage.

Strategic planning for Six Sigma project management is not accomplished merely once. Instead, like any other strategic planning function, it is a cycle of continuous improvements. The improvements can be small or large, measured quantitatively or qualitatively, and designed for either internal or external customers.

There almost always exists a multitude of ideas for continuous improvements. The biggest challenges lie in selecting projects effectively and then assigning the right players. Both of these challenges can be overcome by assigning Six Sigma project management best practices to the project management office. It may even be beneficial to have Six Sigma specialists with green belts or black belts assigned to the PMO.

UNDERSTANDING SIX SIGMA

Six Sigma is not about manufacturing widgets. It is about a focus on processes. And since the PMO is the guardian of the project management processes, it is only fitting that the PMO have some involvement in Six Sigma. The PMO may be more actively involved in identifying the "root cause" of a problem than in managing the Six Sigma solution to the problem.

Some people contend that Six Sigma has fallen short of expectations and certainly does not apply to activities assigned to a PMO. These people argue that Six Sigma is simply a mystique that some believe can solve any problem. In truth, Six Sigma can succeed or fail, but the intent and understanding must be there. Six Sigma gets you closer to the customer, improves productivity, and determines where you can get the biggest returns. Six Sigma is about process improvement, usually repetitive processes, and about reducing the margin for human and/or machine error. The causes of error can only be determined if you understand the critical requirements of either the internal or external customer.

There are a multitude of views and definitions of Six Sigma. Some people view Six Sigma as merely a new name for Total Quality Management (TQM) programs. Others view Six Sigma as a method for implementing a rigorous application of advanced statistical tools throughout the organization. A third view combines the first two views by defining Six Sigma as the application of advanced statistical tools to TQM efforts.

These views are not necessarily incorrect, but rather they are incomplete. From a project management perspective, Six Sigma can be viewed as simply an approach to obtaining better customer satisfaction through continuous process improvement efforts. The customer could be external or internal to the organization, and the word "satisfaction" can have a different meaning depending on whether we are discussing external or internal customers. External customers expect products and services that are a high quality and reasonably priced. Internal customers may define satisfaction in financial terms, such as profit margins. Internal customers may also focus on such items as cycle time reduction, safety requirements, and environmental requirements. If these requirements are met in the most efficient way without any non–value-added costs (i.e., fines, rework, overtime, . . .), then profit margins will increase.

Disconnects can occur between the two definitions of satisfaction. Profits can almost always be increased by lowering quality, but this could jeopardize future business with the client. Making improvements to the methodology to satisfy a particular customer may seem feasible, but may have a detrimental effect on other customers.

The traditional view of Six Sigma focused heavily on manufacturing operations, using quantitative measurements and metrics. Six Sigma tool sets were created specifically for this purpose. Six Sigma activities can be defined as operational Six Sigma and transactional Six Sigma. Operational Six Sigma would encompass the traditional view and focus on manufacturing and measurement. Operational Six Sigma focuses more on processes, such as the enterprise project management methodology, with emphasis on continuous improvements in the use of the accompanying forms, guidelines, checklists, and templates. Some people argue that transactional Six Sigma is merely a subset of operational Six Sigma. While this argument has merit, project management, and specifically the PMO, spend the majority of their time involved in transactional rather than operational Six Sigma.

The ultimate goal of Six Sigma is customer satisfaction but the process by which the goal is achieved can differ depending on whether we are discussing operational or transactional Six Sigma. Table 13–1 identifies some common goals of Six Sigma. The left-hand column lists the traditional goals, which fall more under operational Six Sigma, whereas the right-hand column indicates how the PMO might plan on achieving the goals.

TABLE 13–1　Goals of Six Sigma

Goal[a]	Method of Achievement
• Understand and meet customer requirements (do so through defect prevention and reduction instead of inspection) • Improve productivity	• Improve forms, guidelines, checklists, and templates for understanding customer requirements • Improve efficiency in execution of the project management methodology
• Generate higher net income by lowering operating costs	• Generate higher net income by streamlining the project management methodology without sacrificing quality or performance
• Reduce rework	• Develop guidelines to better understand requirements and minimize scope changes
• Create a predictable, consistent process	• Achieve continuous improvement of the processes

[a]This column was taken from *The Fundamentals of Six Sigma* workbook, p. 1–24. © 2003 by the International Institute for Learning, Inc.; reproduced by permission.

TABLE 13–2   Goals versus Focus Areas

Executive Goals	PMO Focus Areas
• Provide effective status reporting	• Identification of executive needs • Effective utilization of information • "Traffic light" status reporting
• Reduce the time for planning projects	• Sharing information between planning documents • Effective use of software • Use of templates, checklists, and forms
• Improve customer interfacing	• Use of templates for customer status reporting • Completion of customer satisfaction surveys • Extensions of the enterprise project management methodology into the customer's organization

The goals for Six Sigma can be established at either the executive levels or the working levels. The goals may or may not be achievable with the execution of just one project. This is indicated in Table 13–2.

Six Sigma initiatives for project management are not designed to replace ongoing initiatives but to focus on those activities that may have a critical-to-quality and critical-to-customer-satisfaction impact in both the long and short terms.

Operational Six Sigma goals emphasize reducing the margin for human error. But transactional Six Sigma activities managed by the PMO may involve human issues such as aligning personal goals to project goals, developing an equitable reward system for project teams, and project career path opportunities. Fixing people problems are part of transactional Six Sigma but not necessarily part of operational Six Sigma.

SIX SIGMA MYTHS*

There are several myths of Six Sigma, as shown in Table 13–3. These myths have been known for some time but have become quite evident when the PMO takes responsibility for project management of transactional Six Sigma initiatives.

Works Only in Manufacturing
Much of the initial success in applying Six Sigma was based on manufacturing applications; however, recent publications have addressed other applications of

* Adapted from Forrest W. Breyfogle III, James M. Cupello, and Becki Meadows, *Managing Six Sigma.* New York: Wiley, 2001, pp. 6-8. Reproduced by permission of John Wiley & Sons.
† Forrest W. Breyfogle, III, *Implementing Six Sigma; Smarter Solutions Using Statistical Methods.* New York: Wiley, 1999

TABLE 13–3　The Ten Myths of Six Sigma

- Works only in manufacturing
- Ignores the customer in search of bottom-line benefits
- Creates a parallel organization
- Requires massive training
- Is an add-on effort
- Requires large teams
- Creates bureaucracy
- Is just another quality program
- Requires complicated, difficult statistics
- Is not cost-effective

Six Sigma, for example Breyfogle, in *Implementing Six Sigma,*† includes many transactional/service applications. In GE's *1997 Annual Report,* CEO Jack Welch proudly stated that Six Sigma "focuses on moving every process that touches our customers—every product and *service* [emphasis added]—toward near-perfect quality."

Ignores the Customer in Search of Profits

This statement is not myth, but rather misinterpretation. Projects worthy of Six Sigma investments should (1) be of primary concern to the customer, and (2) have the potential for significantly improving the bottom line. Both criteria must be met. The customer is driving this boat. In today's competitive environment, there is no surer way of going out of business than ignoring the customer in a blind search for profits.

Creates a Parallel Organization

An objective of Six Sigma is to eliminate every ounce of organizational waste that can be found and then reinvest a small percentage of those savings to continue priming the pump for improvements. With the large amount of downsizing that has taken place throughout the world during the past decade, there is no room or inclination to waste money through the duplication of functions. Many functions are understaffed as it is. Six Sigma is about nurturing any function that adds significant value to the customer while adding significant revenue to the bottom line.

Requires Massive Training

> Valuable innovations are the positive result of this age [we live in], but the cost is likely to be continuing system disturbances owing to members' nonstop tinkering. . . . [P]ermanent white water conditions are regularly taking us all out of our comfort zones and asking things of us that we never imagined would be required. . . . It is well for us to pause and think carefully about the idea of being

continually catapulted back into the beginner mode, for that is the real meaning of being a continual learner. . . . We do not need competency skills for this life. We need incompetency skills, the skills of being effective beginners. (Vaill)*

Is an Add-On Effort

This is simply the myth "creates a parallel organization" in disguise. Same question, same response.

Requires Large Teams

There are many books and articles within business literature declaring that teams have to be small if they are to be effective. If teams are too large, the thinking goes, a combinational explosion occurs in the number of possible communication channels between team members, and hence no one knows what the other person is doing.

Creates Bureaucracy

A dictionary definition of bureaucracy is "Rigid adherence to administrative routine." The only thing rigid about wisely applied Six Sigma methodology is its relentless insistence that customer needs be addressed.

Is Just Another Quality Program

Based upon the poor performance of untold quality programs during the past three to five decades (Micklethwait and Wooldridge, 1997),† an *effective* quality program would be welcome. More to the point (Pyzdek, 1999),‡ Six Sigma is "an entirely new way to manage an organization."

Requires Complicated, Difficult Statistics

There is no question that a number of advanced statistical tools are extremely valuable in identifying and solving process problems. We believe that practitioners need to possess an analytical background and understand the wise use of these tools, but do not need to understand all the mathematics behind the statistical techniques. The wise application of statistical techniques can be accomplished through the use of statistical analysis software.

---

* Peter B. Vaill, original source unknown.
† John Micklethwait, and Adrian Wooldridge, *The Witch Doctors of the Management Gurus.* New York: Random House, 1997
‡ Thomas Pyzdek, "Six Sigma Is Primarily a Management Program," *Quality Digest,* 1999, p. 26.

Is Not Cost-Effective

If Six Sigma is implemented wisely, organizations can obtain a very high rate of return on their investment within the first year.

USE OF ASSESSMENTS ────────────────────

One of the responsibilities that can be assigned to a PMO is the portfolio management of projects. Ideas for potential projects can originate anywhere in the organization. However, ideas specifically designated as transactional Six Sigma projects may need to be sought out by the PMO.

One way to determine potential projects is through an assessment. An assessment is a set of guidelines or procedures that allows an organization to make decisions about improvements, resource allocations, and even priorities. Assessments are ways to:

- Examine, define, and possibly measure performance opportunities
- Identify knowledge and skills necessary for achieving organizational goals and objectives
- Examine and solve performance gap issues
- Track improvements for validation purposes

A gap is the difference between what currently exists and what should exist. A gaps can involve cost, time, quality, and performance or efficiency. Assessments allow us to pinpoint the gap and determine what knowledge, skills, and abilities necessary to lessen, if not close, the gap. For project management gaps, the assessments can be heavily biased toward transactional rather than operational issues, and this could easily result in behavior-modification projects.

There are several factors that must be considered prior to performing an assessment. These factors might include:

- Amount of executive level support and sponsorship
- Amount of line management support
- Focus on broad-based applications
- Determining who to assess
- Bias of the participants
- Reality of the answers
- Willingness to accept the results
- Impact on internal politics

The purpose of the assessment is to identify ways to improve global business practices first, and functional business practices second. Because the target audience is usually global, there must exist unified support and understanding of the assessment process and of the fact that it is in the best interest of the entire organization. Politics, power, and authority issues must be put aside for the betterment of the organization.

Assessments can take place at any level of the organization. These can be:

- Global organizational assessments
- Business unit organizational assessments
- Process assessments
- Individual or job assessments
- Customer feedback assessments (satisfaction and improvements)

There are several tools available for assessments. A typical list might include:

- Interviews
- Focus groups
- Observations
- Process maps

Assessments for Six Sigma project management should not be performed unless the organization believes that opportunities exist. The amount of time and effort expended can be significant, as shown in Figure 13–3.

The advantages of assessment are that it can lead to significant improvements in customer satisfaction and profitability. However, there are also disadvantages, such as:

- Costly process
- Labor-intensive
- Difficulty in measuring which project management activities can benefit from assessments
- May not provide any meaningful benefits
- Cannot measure a return on investment from assessments

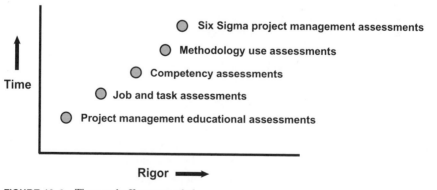

FIGURE 13–3. Time and effort expended.

Assessments can have a life of their own. There are typical life cycle phases for assessments. These life cycle phases may not be aligned with the life cycle phases of the enterprise project management methodology and may be accomplished more informally than formally. Typical assessment life cycle phases include:

- Gap or problem recognition
- Development of the appropriate assessment tool set
- Conducting the assessment/investigation
- Data analyses
- Implementation of the changes necessary
- Review for possible inclusion in the best practices library

Determining the tool set can be difficult. The most common element of a tool set is a focus on questions. Types of questions include:

- Open-ended:
  - Sequential segments
  - Length
  - Complexity
  - Time needed to respond
- Closed-ended:
  - Multiple choice
  - Forced choices (yes–no, true–false)
  - Scales

Table 13–4 illustrates how scales can be set up. The left-hand column solicits a qualitative response and may be subjective, whereas the right-hand column would be a quantitative response and more objective.

It is vitally important that the assessment instrument undergo pilot testing. The importance of pilot testing would be to:

- Validate understanding of the instructions
- Evaluate ease of response
- Determines how much time is needed to respond
- Make sure there is enough space to respond
- Analyze any bad questions

TABLE 13–4   Scales

• Strongly agree	• Under 20%
• Agree	• Between 20 and 40%
• Undecided	• Between 40 and 60%
• Disagree	• Between 60 and 80%
• Strongly disagree	• Over 80%

PROJECT SELECTION

Six Sigma project management focuses on continuous improvements to the enterprise project management methodology. Identifying potential projects for the portfolio is significantly easier than getting them accomplished. There are two primary reasons for this:

- Typical PMOs may have no more than three or four employees. Based upon the activities assigned to the PMO, the employees may be limited as to how much time they can allocate to Six Sigma project management activities
- If functional resources are required, then the resources may be assigned first to those activities that are mandatory for the ongoing business of the firm

The conflict between ongoing business and continuous improvements occurs frequently. Figure 13–4 illustrates this point. The ideal Six Sigma project management activity would yield high customer satisfaction, high cost-reduction opportunities, and significant support for the ongoing business. Unfortunately, what is in the best interest of the PMO may not be in the best, near-term interest of the ongoing business.

All ideas, no matter how good or how bad, are stored in the "idea bank." The ideas can originate from anywhere in the organization, namely:

- Executives
- Corporate Six Sigma champions
- Project Six Sigma champions
- Master black belts

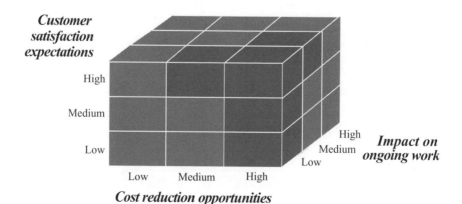

FIGURE 13–4. Project selection cube.

- Black belts
- Green belts
- Team members

If the PMO is actively involved in the portfolio management of projects, then the PMO must perform feasibility studies and cost/benefit analyses on projects together with prioritization recommendations. Typical opportunities can be determined using Figure 13–5. In this figure, ΔX represents the amount of money (or additional money) being spent. This is the input to the evaluation process. The output is the improvement, ΔY, which represents the benefits received or cost savings realized. As an example, consider the following situation.

Situation: Convex Corporation

Convex Corporation identified a possible Six Sigma project involving the streamlining of internal status reporting. The intent was to eliminate as much paper as possible from the bulky status reports and replace them as much as possible with color-coded "traffic light" reporting using the company intranet. The PMO used the following data:

- Burdened hour at the executive level = $240
- Typical number of project status review meetings per project = 8
- Duration per meeting = 2 hours
- Number of executives per meeting = 5
- Number of projects requiring executive review = 20

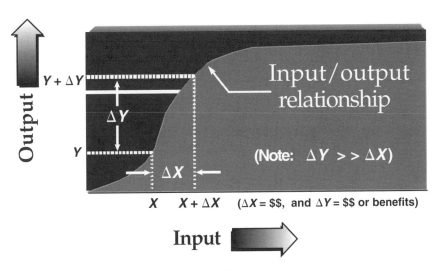

FIGURE 13–5. Six Sigma quantitative evaluation.

Using the above information, the PMO calculated the total cost of executives as: (8 meetings) × (5 executives) × (2 hr/meeting) × ($240/hour) × (20 projects) = $384,000. Convex assigned one systems programmer (burdened at $100/hr) for four weeks. The cost for adding traffic light reporting to the intranet methodology was $16,000.

Six months after implementation, the number of meetings had been reduced to five per project, with an average duration of thirty minutes. The executives were now focusing on only those elements of the project that were color-coded as a potential problem. On a yearly basis, the cost for the meetings on the 20 projects was now about $60,000. In the first year alone, the company identified a savings of $324,000 for one investment of $16,000.

TYPICAL PMO SIX SIGMA PROJECTS

Projects assigned to the PMO can be operational or transactional but mainly the latter. Typical projects might include:

- **Enhanced status reporting:** This project could utilize traffic light reporting designed to make it easier for customers to analyze performance. This could be intranet-based. The intent is to achieve paperless project management. The colors could be assigned based upon problems, present or future risks, or title, level, and rank of the audience.
- **Use of forms:** The forms should be user-friendly and easy to complete. Minimal input by the user should be required, and the data inputted into one form should service multiple forms if necessary. Nonessential data should be eliminated. The forms should be cross-listed to the best practices library.
- **Use of checklists/templates:** These documents should be comprehensive yet easy to understand. They should be user-friendly and easy to update. The forms should be flexible such that they can be adapted to all situations.
- **Definition of success (and failure):** There must exist an established criteria for what constitutes success or failure on a project. There must also exist a process that allows for continuous measurement against this criteria as well as a means by which success (or failure) can be redefined.
- **Team empowerment:** This project looks at the use of integrated project teams, the selection of team members, and the criteria to be used for evaluating team performance. This project is designed to make it easier for senior management to empower teams.
- **Alignment of goals:** Most people have personal goals that may not be aligned with goals of the business. This topic includes project versus company goals, project versus functional goals, project versus individual goals, project versus professional goals, and other such alignments. The

greater the alignment between goals, the greater the opportunity for increased efficiency and effectiveness.

- **Measuring team performance:** This project focuses on ways to uniformly apply critical success factors and key performance indicators to team performance metrics. This also includes the alignment of performance with goals and rewards with goals. This project may interface with the wage and salary administration program by requiring two-way and three-way performance reviews.

- **Competency models:** Project management job descriptions are being replaced with competency models. A competency criteria must be established, including goal alignment and measurement.

- **Financial review accuracy:** This type of project looks for ways of including the most accurate data into project financial reviews. This could include transferring data from various information systems such as earned value measurement and cost accounting.

- **Test failure resolution:** Some PMOs maintain a failure reporting information system that interfaces with failure mode and effect analysis (FMEA). Unfortunately, failures are identified but there may be no resolution on the failure. This project attempts to alleviate this problem.

- **Preparing transitional checklists:** This type of project is designed to focus on transition or readiness of one functional area to accept its responsibility. As an example, it may be possible to develop a checklist on evaluating the risks or readiness of transitioning the project from engineering to manufacturing. The ideal situation would be to develop one checklist for all projects.

The list above is by no means comprehensive. However, the list does identify typical projects managed by the PMO. Some conclusions can be reached by analyzing this list. First, the projects can be both transactional and operational. Second, the majority of the projects focus on improvements to the methodology. Third, having people with Six Sigma experience (i.e. green, brown, or black belts) would be helpful.

When a PMO takes the initiative in Six Sigma project management, then the PMO may develop a Six Sigma toolbox exclusively for the PMO. These tools most likely will not include the advanced statistics tools that are used by the black belts in manufacturing, but may be more process oriented tools or assessment tools.

How to Conduct a Project Management Maturity Assessment*

INTRODUCTION

Once you've decided that a PM maturity assessment is right for your organization, you'll be faced with the somewhat daunting realization that now you've got to actually plan, organize, and implement the assessment. Where do you start? And how do you turn all that assessment data into a meaningful action plan? In this chapter you will be provided with some helpful guidelines and checklists.

FIND WAYS TO BYPASS THE CORPORATE IMMUNE SYSTEM

Even though what you're doing is for the good of the organization, you may encounter cultural resistance as you prepare to implement a PM maturity assessment. That's because any organization is like a biological organism. It will tend to reject anything that is new and unfamiliar (like the body's immune system rejecting a life-saving transplant). Intellectually the organization will appreciate the

* This chapter has been adapted from Harold Kerzner, *Advanced Project Management: Best Practices in Implementation.* New York: Wiley, 2004, pp. 197–208; reproduced by permission of John Wiley & Sons. This chapter was originally prepared by G. Howland Blackiston, formerly Executive Vice President, International Institute for Learning, Inc. For further information on the Project Management Maturity Assessment Instrument referenced in this chapter, contact Lori Milhaven, Vice President, International Institute for Learning, at 212-515-5121 or through e-mail: lori.milhaven@iil.com

value of the assessment, but the company's culture can be a troublemaker. So be sure to take specific steps to prevent "rejection" and to ensure success.

- Recognize and anticipate that there will be pockets of resistance
- Acknowledge the fear factor: the apprehension that we may be doing things wrong.
- Identify the specific cultural issues that might cause resistance. Are there personal issues involved (concern about status or job security)?
- Are there legal restrictions regarding asking employees to take an assessment? Some countries have laws on the books that make assessments difficult.
- Squarely address each and every concern. Defuse resistance by acknowledging problems. Be honest and candid.
- Begin an assessment using volunteers who share your enthusiasm. Let their positive experiences convince the others.
- Launch an assessment effort by beginning with a part of the business that is project driven (e.g., IT, R&D, marketing)
- Start small and scale up. Learn from your early successes and failures before you launch a company-wide assessment effort. Walk before you run.
- Clearly communicate exactly what you are doing and why (see the next section)

EXPLAIN WHY YOU ARE DOING THIS

A well-planned and executed PM maturity assessment is time consuming. It will involve careful planning and follow-up, and will draw on considerable resources. So be sure there's clear understanding in the organization as to why you are doing this. Here's where good communication skills come in handy. Prepare a brief and lucid document that can be shared with everyone who will participate in the PM maturity assessment. You've got to "sell" the importance of doing this assessment. And you've got to disarm resistance. This will help ensure buy-in and head off any problems that can derail success. As you put together this memo, make sure you address the following issues:

- Define what is meant by "project management maturity."
- Explain why it's important for your company to measure PM maturity.
- Convey how this assessment will make the organization more competitive.
- Underscore how competitive companies create growth and job security.
- Explain who in the company will be invited to participate in the assessment. Why are these people being chosen?
- Describe what is involved and how long it will take.
- Recognize what management will do (and not do) with the assessment information.
- Alleviate any concerns that the information will be used to judge an individual's performance (don't threaten job security).

- Explain how the organization will turn the assessment data into specific improvements.
- Communicate any plans for doing this assessment again (you will want to measure ongoing progress).

Create an Effective Welcome Message

(Here's a sample welcome message from a company using the Kerzner Project Management Maturity Online Assessment Tool. This message appeared on the home page of the web-based assessment tool. Clear statements like this help alleviate cultural concerns about participating in what might be incorrectly regarded as a threatening "test of knowledge".)

Welcome to our online project management maturity assessment tool. Project management has been recognized as playing an essential role in our organization. By participating in this assessment you will help create a strategic development plan for identifying the training curriculum that will build and improve our current capabilities. Results of this assessment will be used to set a baseline for all departments and will serve as a tool for identifying future training opportunities. Your support demonstrates your commitment to helping our project management community achieve professional recognition, continuous improvement and productivity. In turn, this will result in higher consultant/partner/customer satisfaction. We wish to thank you for participating in this groundbreaking event and helping us realize the company's future vision. Thank you for your participation.

———Corporate Project Office

PICK THE MODEL THAT IS BEST FOR YOUR ORGANIZATION

There's no lack of assessment tools on the market—there are a lot to choose—simple or complex, generic or industry specific. Basically they all seek similar objectives: to measure an organization's project management strengths and weaknesses, and to identify improvement opportunities. No one model will be 100 percent perfect for your organization, but some may come close. In all likelihood you'll wind up with a blend or a customization that best fits your organization. As you evaluate assessment models, consider the following:

- Is it compatible with your project management methodology?
- Does it speak your language (use similar terms and definitions)?

- Has the assessment model been validated (has it been tested and used successfully in other industries)?
- Will this tool work well for your industry? In your organization?
- How easy is the tool to use and administer?
- What delivery mechanism would be best (printed surveys, interviews, online)?
- Is the tool aligned with industry standards [e.g., the Project Management Book of Knowledge (PMBOK®) Guide)?
- Global organizations should determine if the tool is applicable internationally.
- Can the results of the assessment be easily mapped to your organization's business plan?
- Is the tool flexible? Does it allow for special features and customization?
- Can the tool measure professional skills of project personnel?
- What resources will be required to utilize the tool? How many employees will be involved and how long will it take?
- What will it cost to undertake the assessment?

MATURITY MODELS: HOW DO THEY COMPARE? _____

Are you ready to embark upon an assessment? In the spirit of comparative analysis, there are many maturity models currently available other than the model discussed in this book.

The Origin of Today's Maturity Models

Back in the mid to late 1980s, the software industry explored formal ways to better measure and manage the quality and reliability of the processes used for software development. The industry saw value in applying the concepts of Total Quality Management (TQM) and continuous improvement to their development processes. This prompted the Software Engineering Institute's (SEI) development (in 1990) of the Capability Maturity Model (CMM®). The tool provided the industry with a structured and objective means for measuring a software organization's development processes, and comparing these measures against optimum practices. CMM helped software developers identify specific improvements that would allow them to become more competitive in a highly competitive industry. To utilize CMM in other industries, the tools have been blended with project management measures and standards (à la the PMBOK Guide) to serve as the foundation for many of the project management maturity models now on the market.

CREATE THE RIGHT FIT ——————————————————

Some models (like the Kerzner Project Management Maturity Model) have been designed to meet the needs of a broad array of industries and cultures. They are generic. Other models have been developed for specific industries or applications. As you select a model to use in your organization, consider to what degree (if any) the model must be tailored to fit your culture, industry, and business objectives. Some issues to consider are:

- Is the model based on a project management standard that fits with what's used in your organization? Or will you have to tailor the tool to comply with your standards?
- Does the model work in your industry? Are the terms and language used familiar to your business?
- Is there a cultural fit?
- Is the model comprehensive enough to measure leadership, professional development, and management involvement?
- Will the model help you develop a corrective action plan to continuously improve project management processes and practices?
- Does the model allow you to add questions and make modifications without compromising the effectiveness of the assessment?
- Can you sort assessment results to take into account different roles and responsibilities; various departments; geographic locations; or job functions?

It's OK to Make Changes

When tailoring a maturity model to better fit your organization, making changes to the model is perfectly acceptable. For an example, look at our Kerzner Model. Note that Level Three determines if the organization is using a "singular methodology" (rather than using multiple methodologies). Some organizations may intentionally use several methodologies rather than one—for example, one for IT and another for new product development. By all means tailor the model to fit the realities of your organization.

CHOOSE AN APPROPRIATE DELIVERY METHOD ——————————

As you ponder over how you will "deliver" the assessment instrument to your audience, keep in mind that no one way is correct. There are a number of options available to you. The method you choose may depend upon your audience, size of company, time available, budget, flexibility, and even technology. Regardless of the option you choose, it's helpful to clarify the time frame for completing the assessment. Let your audience know when you need it back. Don't give them too much time or they'll put it aside forever. Tell them you need the completed assessment within a couple of weeks. That way you'll have a better chance of getting what you need in a timely way.

Here are some things to consider:

- Decide if you want an informal or a formal approach. If your organization is small and straightforward, an informal assessment may be all you need.
- If you decide to use conventional questionnaires (paperwork), keep in mind the logistics of distributing, collecting, and tabulating all the data.
- Consider conducting interviews to gather the data you need. This involves some tricky scheduling issues, but if the numbers are manageable, this might be a doable option.
- Don't overlook the possibility of utilizing online technology. This is a very convenient way to reach a large audience in a short time. In addition, the tabulation of results is automatic and instantaneous. And the online format permits easy editing and modification of the tool itself (see sidebar).
- Pick a model and stick with it. Using different instruments may confound your ability to take meaningful corrective action.
- How about using an outside resource (e.g., consultants)? Their objective approach can add value to the assessment results. Often reports from impartial, outside experts are more readily accepted than the same reports from one of your own.

Use Online Technology to Your Advantage

Creating an interactive web-based assessment has its advantages: easily editable; auto-scoring; efficient distribution, and so on. A few things to consider are covered in the following, along with some screen shots of the Kerzner Project Management Maturity Online Assessment Tool to illustrate how it works. If you decide to spend the time and resources to develop your own web-based assessment, make sure you take into account the following:

- Create an interface that is intuitive and easy to use.
- Make the scoring function automatic.
- Provide for an auto-sorting of results by critical filters (key departments; divisions; job functions; hierarchy; etc.).
- Build an "Executive Dashboard" feature so that the Project Office and/or top management can continuously monitor the assessment results.

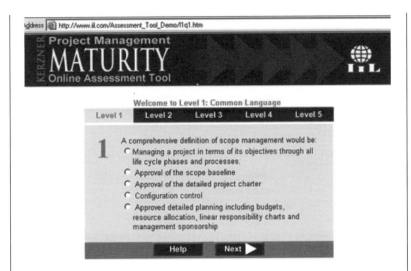

Participants using the Kerzner Assessment Tool answer a series of multiple-choice questions within each of the model's five levels. There are a total of 183 questions in the Kerzner assessment tool.

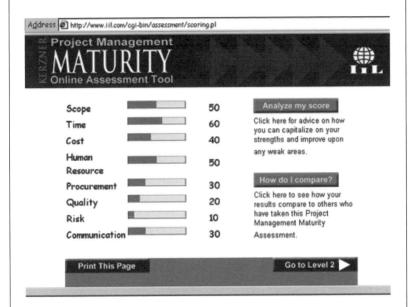

When completing each level of the Kerzner assessment, participants can see their own score results broken down by the subcategories within that level. This helps them recognize specific strengths and weaknesses.

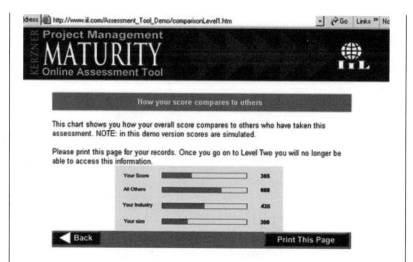

A click of a button compares an individual's score to all others who have taken the Kerzner assessment—both inside and outside of their organization (kind of a sanity check to see how the results compare to the rest of the world). Users can also see how their company's overall scores compare to other companies in similar industries. Scores are automatically broken down by whatever filtering criteria have been established by the organization. The tool also allows users to display a prescriptive narrative analysis of their results. This auto-generated report suggests what can be done to advance to higher levels of project management maturity (the suggestions are stimulated by the assessment scores).

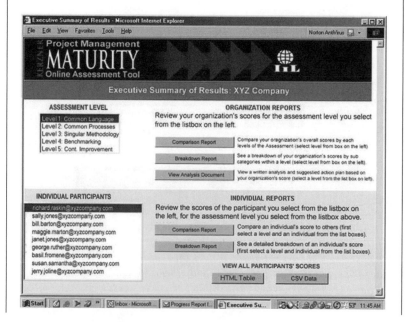

> *The Kerzner Online Assessment Tool allows authorized managers to see a live summary of their organization's results. Managers can view and compare both individual and company-wide assessment scores. They can even export the raw data into other applications (such as Excel).*

ESTABLISH RESPONSIBILITY

In the best of all worlds, the Project Office (assuming you have one) should assume overall responsibility for the project management maturity assessment. Chances are this office reports directly to executive management. Or perhaps its membership is comprised of executive management. In either case, that link to top management will come in handy, because the assessment will identify opportunities for improvement that will rely on decisions and directives that only they can supply:

- Providing overall leadership for the assessment
- Helping with the selection/design of the assessment tool
- Identifying who should participate in the assessment
- Monitoring assessment results and aligning the resulting improvement opportunities with the organization's business plan
- Setting priorities (identifying which few actions will have the most meaningful impact)
- Developing an action plan that will allow the organization to achieve these improvements
- Supplying the needed resources (time, budget, personnel)
- Approving the remedial training curriculum
- Encouraging a broadening of the assessment throughout the organization
- Evaluating ongoing progress

DECIDE WHO SHOULD PARTICIPATE

Here's where you need to make some strategic decisions. Who should participate in the assessment? Ideally, you should include a broad and diverse representation of the entire company—the broader the representation, the more objective and accurate the assessment. Assuming you are attempting to make the company more competitive in project management, you should be assessing the entire organization's project management maturity (not merely assessing the maturity of a single department).

Getting everyone involved may be your ultimate goal, but it may be easier on the company culture to begin in a receptive department before scaling up to include the entire organization. In either case, you want to be certain that you have the right people participating in the assessment. Consider the following guidelines as you prepare your invitation list:

- Use the company's business plan as a guide to identify which areas of the organization will critically rely on project management expertise. These departments, locations, or functions are obvious candidates for participating in the assessment.
- Get the right cross-representation involved (the right composition is critical).
- You will underutilize the assessment if you limit participation to experienced project managers (this narrow approach will not accurately reflect the company's actual project management maturity—it's a biased assessment of reality).
- Be sure to include key customers and stakeholders.
- Include broad representation across departments (remember that in world-class companies, all departments, functions, and levels are involved in or knowledgeable about project management efforts).
- Include representation of the entire hierarchy (executive; middle; lower; associates).
- Include a large enough sample to be statistically valid. You should invite at least 30 or more individuals from each operating unit each time you conduct the assessment.
- For large companies, your ultimate objective is to assess at least 10 percent of the population (target larger percentages in small companies).
- Create the ability to sort assessment results by key departments/divisions/ job functions/hierarchy (such sorting will enhance your ability to identify project management strengths and weaknesses).
- Decide if participation should be mandatory or optional. For cultural acceptance, I suggest that participation is optional at the start, and mandatory as you scale up your assessment efforts and are able to demonstrate value (see "Find Ways to Bypass the Corporate Immune System").

TURN THE RESULTS INTO AN ACTION PLAN

With the assessment results in hand, the data should be used to identify meaningful improvement activities. You've got to turn scores into corrective action plans. The prioritization and deployment of these improvements should be spearheaded by the company's Project Office. If no such office is in place, consider creating a PMM assessment action team. Any significant effort will ultimately require the support of top management.

As you organize this effort, keep the following in mind:

- Consider using an objective outside resource to interpret and analyze the assessment results (they've got no axe to grind, and often recommendations from outsiders are more readily accepted by the culture)

- Be as specific as possible when converting assessment data into specific improvement actions. There's little value in concluding: "Parts of the company need to improve their understanding of risk management." Be much more specific. *How* is the company going to achieve this goal? A better action would be "Starting in December, we will schedule two-day intensive workshops in 'Risk Management' for key personnel in our marketing and legal departments."
- Don't forget to identify and deal with any obstacles in the way of making improvements.
- Treat each improvement objective as an individual improvement project (keep in mind that *all* improvement takes place *project-by-project,* and in no other way).
- When prioritizing improvements, start with some home runs (create those bellwether examples that help sell the value of what's being done).
- Focus first on those things that improve the business (prioritize actions based on the organization's business goals and objectives). Use Pareto Analysis to identify the vital few improvements that will have the greatest positive impact on the company.
- Use the assessment tool repeatedly to measure progress. Companies that are serious about improving project management should conduct project management maturity assessments at least once every quarter.

DEVELOP A REMEDIAL TRAINING CURRICULUM ⎯⎯⎯⎯⎯

The assessment will be helpful in identifying a training curriculum that will help you "close the gaps." The analysis of scores will clearly identify where training is needed, and in what subjects.

There is great economy here. It means you only need to train those individuals, functions, departments, or locations that have been identified as needing training. It also means you only need training in those specific subjects that have been identified as lacking. This subscribes to the concept of "just in time" training, versus "just in case" training.

The Project Office (or top management) should establish a task force to plan the organization's remedial training curriculum. Remedial training should be mandated (not voluntary). This group should ensure the following:

- Make certain the assessment results have been analyzed in such a way that it's clear what must be done to make meaningful improvements. If necessary, hire outside experts to help you with the analysis.
- Establish the criteria to be followed in designing or selecting the curriculum.
- Keep the training curriculum keenly responsive to the subject needs and target audiences identified by the assessment ("just in time" versus "just in case").
- Decide whether to buy training from outside firms or develop your own.

- Prepare local case materials as supplements to training. This keeps the tools and methodologies relevant to your business. It also helps participants understand how to apply their new knowledge to their jobs.
- Adapt interactive exercises to fit your culture and job situations.

KEEP TOP MANAGEMENT INFORMED

The overall purpose of doing a project management maturity assessment is to identify opportunities for making significant improvements in the way you manage projects. In turn such improvements lead to better project outcomes, lower costs, faster results, higher quality and greater customer satisfaction. But no significant corrective action is possible without the involvement of top management. They are the only ones who can authorize the significant time and resources needed to turn the assessment results into a specific action plan. Keeping top management informed is vital if you are to win their support and leadership.

Part of your assessment plan should include a means for keeping top management informed. I know of one company that has 100 employees participate in a project management maturity assessment every month. And every month top management is given a report summarizing the assessment results. This flow of information helps management prioritize action plans, identify training needs, and measure the company's improvement progress.

Keep the reporting relevant to top management's needs. Share information that is most meaningful to them. This will vary from company to company, but you can be sure that anything expressed in the "language of money" will get their attention and stimulate action. Here are some suggestions:

- Present a detailed breakdown of scores to clearly identify the company's specific strengths and weaknesses.
- Show a comparison of scores between different departments, job functions, geographic locations, or whatever filtering criteria are most meaningful to your business objectives.
- You may wish to provide management with a breakdown of scores by individuals participating in the assessment. This information can be used to identify outstanding talent. And it can even be used to identify individuals who would benefit from remedial training. But be careful! If this information is used to reprimand, criticize, or "clean house", you will effectively crush cultural acceptance of the assessment and all will come to a grinding and hopeless halt.
- Prepare an action plan based on the assessment results. Be specific as to what corrective action is needed and why (see the next section of this chapter).

VIRTUAL REPORTING _____

For those utilizing an "online" assessment tool, a helpful option is to create a "on-line" reporting feature. I refer to this as an "Executive Dashboard." This consists of a unique URL address that allows authorized managers to see a detailed summary of their organization's assessment results. Because it's online, the information is real-time, displaying the latest, up-to-date scores each time it is accessed. Instant gratification! Managers can view and compare individual and company-wide assessment scores whenever they wish. The feature can also allow them to export the raw data into spreadsheet applications (for other reports, sorting options, etc.). Keep the interface simple and intuitive to use.

BENCHMARK YOUR RESULTS TO OTHERS _____

It's helpful to compare your results with those of others who have taken the same assessment (note that the online version of the Kerzner PM Maturity Assessment has this feature). Such comparisons should be both within and outside of your industry. Benchmarking results is helpful for the following reasons:

- Provides a "sanity check" within your industry—is your maturity level close to your competitor's?
- Gives you a realistic target, proving that achieving higher levels of maturity is possible (after all, others have already reached higher levels).
- Avoids the deadly sin of "resting on your laurels" (if you are complacent, assuming you're already best within your industry, it's a sobering jolt to see that other industries are actually much better than you are).
- Sells the need and urgency for improvement (management will be motivated to action if they see that other companies are outperforming your organization).

DO IT AGAIN _____

As helpful as a maturity assessment can be, its usefulness is minimized when it's regarded as a one-time event. That's an underutilization of a powerful tool. Sure you'll be able to identify your strengths, weaknesses, and opportunities for improvement—that will help you develop a corrective action plan. But it's only when you use the assessment on a *repetitive* basis that you can objectively measure the progress of your corrective action plan.

- Are your overall scores improving?
- Is the company achieving higher levels of project management maturity?

- How do you compare to your competition?
- Do you need to modify your corrective action plan, based on the latest assessment results?
- Have new opportunities for improvement emerged since the last assessment?
- Can you improve the assessment tool for more effective use in our organization?
- If one division is outperforming others, are there skills and methods within that exemplary division that can be applied elsewhere in the organization?

Stay nimble and in tune with the marketplace by conducting the assessment on a regular basis. Ongoing use of the tool also allows you to evaluate a larger and larger percentage of your total population. How often should you conduct the assessment? That depends on your organization. Here are some guidelines:

- If your organization is project driven (projects are critical to your business success), you should perform an assessment every month. Vary the audience each time, striving to capture a broad and diverse representation of the organization.
- Other organizations should conduct the assessment a minimum of once a quarter to ensure that improvements are being made. Again, vary the audience each time, striving to capture a broad and diverse representation of the organization.
- Always keep top management informed.

Understanding Best Practices

INTRODUCTION

One of the benefits of using the project management maturity model (PMMM) is that it provides you with a means for capturing and retaining best practices. Unfortunately this is easier said than done. There are multiple definitions of a best practice such as:

- Something that works
- Something that works well
- Something that works well on a repetitive basis
- Something that leads to a competitive advantage
- Something that can be identified in a proposal to generate business

In the author's opinion, *best practices are those actions or activities under-taken by the company or individuals that lead to a sustained competitive advantage in project management.*

It has only been in recent years that the importance of best practices has been recognized. In the early years of project management, there were misconceptions concerning project management. Some of the misconceptions included the ideas that:

- Project management is a scheduling tool similar to PERT/CPM scheduling.
- Project management applies to large projects only.
- Project management is designed for government projects only.

- Project managers must be engineers and preferably with advanced degrees.
- Project managers need a "command of technology" to be successful.
- Project success is measured in technical terms only.

As project management evolved, best practices became important. Best practices can be learned from both successes and failures. In the early years of project management, private industry focused on learning best practices from successes. The government, however, focused on learning about best practices from failures. When the government finally focused on learning from successes, the knowledge on best practices came from their relationships with both their prime contractors and the subcontractors. Some of the best practices that came out of the government included:

- Use of life cycle phases
- Standardization and consistency
- Use of templates [i.e., for statement of work (SOW), work breakdown structure (WBS), risk management, etc.]
- Providing military personnel in project management positions with extended tours of duty at the same location
- Use of integrated project teams (IPT)
- Control of contractor-generated scope changes
- Use of earned value measurement

WHAT TO DO WITH A BEST PRACTICE

If a best practice is defined as one that leads to a sustained competitive advantage, it is no wonder that some companies would be reluctant to make their best practices known to the general public. Therefore, what should a company do with its best practices if not publicize them? The most common options available include:

- **Sharing knowledge internally only:** This is accomplished using the company intranet to share information with employees. There may be a separate group within the company responsible for control of the information, perhaps even the Project Management Office (PMO).
- **Hiding knowledge from all but a selected few:** Some companies spend vast amounts of money on the preparation of forms, guidelines, templates, and checklists for project management. These documents are viewed as both company-proprietary information and best practices, and are provided to only a select few on a need-to-know basis. An example of a "restricted" best practice might be specialized forms and templates for project approval wherein information contained within may be company-sensitive financial data, or the company's position on profitability and market share.

● **Advertising to your customers:** In this approach, companies may develop a best practices brochure to market their achievements and may also maintain an extensive best practices library that is shared with their customers—but only after contract award.

Even though companies collect best practices, not all best practices are shared outside of the company even during benchmarking studies where all parties are expected to share information. Students often ask why textbooks do not include more information on detailed best practices such as forms and templates. One company commented to the author,

> We must have spent at least $1 million over the last several years developing an extensive template on how to evaluate the risks associated with transitioning a project from engineering to manufacturing. Our company would not be happy giving this template to everyone who wants to purchase a book for $80. Some best practices templates are common knowledge and we would certainly share this information. But we view the transitioning template as proprietary knowledge not to be shared.

CRITICAL QUESTIONS

There are several questions that must be addressed before an activity is recognized as a best practice. Three frequently asked questions include:

● Who determines that an activity is a best practice?
● How do you properly evaluate what you think is best practice to validate that in fact it is a true best practice?
● How do you get executives to recognize that best practices are true value-added activities and should be championed by executive management?

Some organizations have committees that have as their primary function the evaluation of potential best practices. Other organizations use the PMO to perform this work. These committees most often report to the senior levels of management. Evaluating whether or not something is a best practice is not time-consuming but it is complex. Simply because someone believes that what they are doing is a best practice does not mean that it is in fact a best practice. PMOs are currently developing templates and criteria for determining that an activity may qualify as a best practice. Some criteria that are included in the template might be:

● Transferable to many projects
● Enables efficient and effective performance that can be measured
● Enables measurement of possible profitability using the best practice
● Allows an activity to be completed in less time and at a lower cost

One company had two unique characteristics in their best practices template:

● The best practice helps to avoid failure.
● If a crisis develops, the best practice helps us to get out of a critical situation.

Executives must realize that these best practices are, in fact, intellectual property to benefit the entire organization. If the benefits of a best practice can be quantified, then it is usually easier to convince senior management of its value.

LEVELS OF BEST PRACTICES

Best practices can be discovered anywhere within or outside of your organization. Figure 15–1 shows various levels of best practices. The bottom level is the professional standards level, which would include professional standards as defined by the Project Management Institute (PMI®). The professional standards level contains the greatest number of best practices, but they are more of a general nature than specific, and have a low level of complexity.

The industry standards level would identify best practices related to performance within the industry. For example, the automotive industry has established standards and best practices specific to the auto industry.

As we progress from the professional standards to the individual best practices in Figure 15–1, the complexity of the best practices increases and, as ex-

FIGURE 15–1. Levels of best practices.

pected, the quantity of best practices decreases. Examples of best practice at each level (from general to specific) might be:

- **Professional standards:** Preparation and use of a risk management plan including templates, guidelines, forms, and checklists for risk management.
- **Industry-specific:** The risk management plan includes industry best practices such as the best way to transition from engineering to manufacturing.
- **Company-specific:** The risk management plan identifies the roles and interactions of engineering, manufacturing, and quality assurance groups during transition.
- **Project-specific:** The risk management plan identifies the roles and interactions of affected groups as they relate to a specific product/service for a customer.
- **Individual:** The risk management plan identifies the roles and interactions of affected groups based upon their personal tolerance for risk, possibly through the use of a Responsibility Assignment Matrix prepared by the project manager.

Best practices can be extremely useful during strategic planning activities. As shown in Figure 15–2, the bottom two levels may be more useful for project strategy formulation, whereas the top three levels are more appropriate for the execution of a strategy.

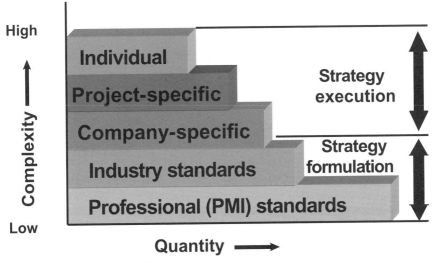

FIGURE 15–2. Usefulness of best practices.

COMMON BELIEFS _____

There are several common beliefs concerning best practices. A partial list includes:

- Because best practices can be interrelated, the identification of one best practice can lead to the discovery of another best practice, especially in the same category or level of best practices.
- Because of the dependencies that can exist between best practices, it is often easier to identify categories of best practices rather than individual best practices.
- Best practices may not be transferable. What works well for one company may not work for another company.
- Even though some best practices seem simplistic and common sense in most companies, the constant reminder and use of these best practices leads to excellence and customer satisfaction.
- Best practices are not limited exclusively to companies in good financial health

Care must be taken that the implementation of a best practice does not lead to detrimental results. One company decided that the organization must recognize project management as a profession in order to maximize performance and retain qualified people. A project management career path was created and integrated into the corporate reward system.

Unfortunately, the company made a severe mistake. Project managers were given significantly larger salary increases than line managers and workers. People became jealous of the project managers and applied for transfer into project management thinking that the "grass was greener." The company's technical prowess diminished and some people resigned when not given the opportunity to become a project manager.

Companies can have the greatest intentions when implementing best practices and yet detrimental results can occur. Table 15–1 identifies some possible

TABLE 15–1   Results of Implementing Best Practices

Type of Best Practice	Expected Advantage	Potential Disadvantage
Use of traffic light reporting	Speed and simplicity	Poor accuracy of information
Use of a risk management template/form	Forward looking and accurate	Inability to see some potential critical risks
Highly detailed WBS	Control, accuracy, and completeness	Excess control and cost of reporting
Using EPM on all projects	Standardization and consistency	Too expensive on certain projects
Using specialized software	Better decision-making	Too much reliance on tools

expectations and the detrimental results that can occur. The poor results could have been the result of poor expectations or of not fully understanding the possible ramifications after implementation.

There are other reasons why best practices can fail or provide unsatisfactory results. These include:

- Lack of stability, clarity, or understanding of the best practice
- Failure to use best practices correctly
- Identifying a best practice that lacks rigor
- Identifying a best practice based upon erroneous judgment

THE BEST PRACTICES LIBRARY

With the premise that project management knowledge and best practices are intellectual property, then how does a company retain this information? The solution is usually the creation of a best practices library. Figure 15–3 shows the three levels of best practices that seem most appropriate for storage in a best practices library.

Figure 15–4 shows the process of creating a best practices library. The bottom level is the discovery and understanding of what is or is not a "potential" best practice. The ideas for potential best practices can originate anywhere within the organization.

FIGURE 15–3. Levels of best practices.

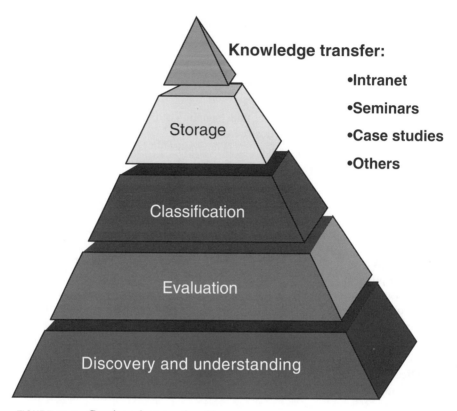

FIGURE 15–4. Creating a best practices library.

The next level is the evaluation level, where the organization tries to confirm that the potential best practice is indeed just that. The evaluation process can be carried out by the PMO or by a committee, but should have involvement by the senior levels of management. The evaluation process is very difficult because a one-time positive occurrence may not reflect a best practice. There must exist established criteria for the evaluation of a best practice.

Once it is agreed upon that a best practice exists, then it must be classified and stored in some retrieval system such as a company intranet best practices library.

Figure 15–1 showed the levels of best practices, but the classification system for storage purposes can be significantly different. Figure 15–5 shows a typical classification system for a best practices library.

The purpose for creating a best practices library is to transfer knowledge to the employees. The knowledge can be transferred through the company intranet, seminars on best practices, and case studies. Some companies require that the project team prepare case studies on lessons learned and best practices before the

Effectiveness and efficiency

Future business

Technology advances

Project management

Satisfaction

FIGURE 15–5. Best practices library.

team is disbanded. These companies then use the case studies in company-sponsored seminars. Best practices and lessons learned must be communicated to the entire organization. The problem is determining how to do it effectively.

Another critical problem is best practices overload. One company started up a best practices library and, after a few years, had amassed what they considered to be hundreds of best practices. For a long time, however, nobody bothered to reevaluate whether or not all of these were still best practices. Once reevaluation finally took place, it was determined that less than one-third of the practices listed were still regarded as best practices. Some were no longer best practices, others needed to be updated, and others had to be replaced with newer best practices.

Packer Telecom

Background

The rapid growth of the telecom industry made it apparent to Packer's executives that risk management must be performed on all development projects. If Packer were late in the introduction of a new product, then market share would be lost. Furthermore, Packer could lose valuable opportunities to "partner" with other companies if Packer were regarded as being behind the learning curve with regard to new product development.

Another problem facing Packer was the amount of money being committed to R&D. Typical companies spend 8 to 10 percent of earnings on R&D, whereas in the telecom industry, the number may be as high as 15 to 18 percent. Packer was spending 20 percent on R&D, and only a small percentage of the projects that started out in the conceptual phase ever reached the commercialization phase, where Packer could expect to recover its R&D costs. Management attributed the problem to a lack of effective risk management.

The Meeting

PM: I have spent a great deal of time trying to benchmark best practices in risk management. I was amazed to find that most companies are in the same boat as us, with very little knowledge in risk management. From the limited results I have found from other companies, I have been able to develop a risk management template for us to use.

Sponsor: I've read over your report and looked at your templates. You have words and expressions in the templates that we don't use here at Packer. This concerns me greatly. Do we have to change the way we manage projects to use these templates?

Are we expected to make major changes to our existing project management methodology?

PM: I was hoping we could use these templates in their existing format. If the other companies are using these templates, then we should also. These templates also have the same probability distributions that other companies are using. I consider these facts equivalent to a validation of the templates.

Sponsor: Shouldn't the templates be tailored to our methodology for managing projects and our life cycle phases? These templates may have undergone validation, but not at Packer. The probability distributions are also based upon someone else's history, not our history. I cannot see anything in your report that talks about the justification of the probabilities.

The final problem I have is that the templates are based upon history. It is my understanding that risk management should be forward looking, with an attempt at predicting the possible future outcomes. I cannot see any of this in your templates.

PM: I understand your concerns, but I don't believe they are a problem. I would prefer to use the next project as a "breakthrough" project using these templates. This will give us a good basis to validate the templates.

Sponsor: I will need to think about your request. I am not sure that we can use these templates without some type of risk management training for our employees.

Questions
1. Can templates be transferred from one company to another, or should tailoring be mandatory?
2. Can probability distributions be transferred from one company to another? If not, then how do we develop a probability distribution?
3. How do you validate a risk management template?
4. Should a risk management template be forward looking?
5. Can employees begin using a risk management template without some form of specialized training?

Case 2

Luxor Technologies

Between 1992 and 1996, Luxor Technologies had seen their business almost quadruple in the wireless communications area. Luxor's success was attributed largely to the strength of its technical community, which was regarded as second to none. The technical community was paid very well and given the freedom to innovate. Even though Luxor's revenue came from manufacturing, Luxor was regarded by Wall Street as being a technology-driven company.

The majority of Luxor's products were based upon low cost, high quality applications of the state-of-the-art technology, rather than advanced state-of-the-art technological breakthroughs. Applications engineering and process improvement were major strengths at Luxor. Luxor possessed patents in technology breakthrough, applications engineering, and even process improvement. Luxor refused to license their technology to other firms, even if the applicant was not a major competitor.

Patent protection and design secrecy were of paramount importance to Luxor. In this regard, Luxor became vertically integrated, manufacturing and assembling all components of their products internally. Only off-the-shelf components were purchased. Luxor believed that if they were to use outside vendors for sensitive component procurement, they would have to release critical and proprietary data to the vendors. Since these vendors most likely also serviced Luxor's competitors, Luxor maintained the approach of vertical integration to maintain secrecy.

Being the market leader technically afforded Luxor certain luxuries. Luxor saw no need for expertise in technical risk management. In cases where the technical community was only able to achieve 75–80 percent of the desired specification limit, the product was released as it stood, accompanied by an announcement that there would be an upgrade the following year to achieve the remaining 20–25 percent of the specification limit, together with other features. Enhancements and upgrades were made on a yearly basis.

Exhibit I. Likelihood of a technical risk

Event	Likelihood Rating
• State-of-the-art advance needed	0.95
• Scientific research required (without advancements)	0.80
• Concept formulation	0.40
• Prototype development	0.20
• Prototype testing	0.15
• Critical performance demonstrated	0.10

By the fall of 1996, however, Luxor's fortunes were diminishing. The competition was catching up quickly, thanks to major technological breakthroughs. Marketing estimated that by 1998, Luxor would be a "follower" rather than a market leader. Luxor realized that something must be done, and quickly.

In January 1999, Luxor hired an expert in risk analysis and risk management to help Luxor assess the potential damage to the firm and to assist in development of a mitigation plan. The consultant reviewed project histories and lessons learned on all projects undertaken from 1992 through 1998. The consultant concluded that the major risk to Luxor would be the technical risk and prepared Exhibits I and II.

Exhibit I shows the likelihood of a technical risk event occurring. The consultant identified the six most common technical risk events that could occur at Luxor over the next several years, based upon the extrapolation of past and present data into the future. Exhibit II shows the impact that a technical risk event could have on each project. Because of the high probability of state-of-the-art advancements needed in the future (i.e., 95 percent from Exhibit I), the consultant identified the impact probabilities in Exhibit II for both with and without state-of-the-art advancement needed.

Exhibits I and II confirmed management's fear that Luxor was in trouble. A strategic decision had to be made concerning the technical risks identified in Exhibit I, specifically the first two risks. The competition had caught up to Luxor in applications engineering and was now surpassing Luxor in patents involving state-of-the-art

Exhibit II. Impact of a technical risk event

Event	Impact Rating	
	With State-of-the-Art Changes	Without State-of-the-Art Changes
• Product performance not at 100 percent of specification	0.95	0.80
• Product performance not at 75–80 percent of specification	0.75	0.30
• Abandonment of project	0.70	0.10
• Need for further enhancements	0.60	0.25
• Reduced profit margins	0.45	0.10
• Potential systems performance degradation	0.20	0.05

advancements. From 1992 to 1998, time was considered as a luxury for the technical community at Luxor. Now time was a serious constraint.

The strategic decision facing management was whether Luxor should struggle to remain a technical leader in wireless communications technology or simply console itself with a future as a "follower." Marketing was given the task of determining the potential impact of a change in strategy from a market leader to a market follower. The following list was prepared and presented to management by marketing:

1. The company's future growth rate will be limited.
2. Luxor will still remain strong in applications engineering but will need to out-source state-of-the-art development work.
3. Luxor will be required to provide outside vendors with proprietary information.
4. Luxor may no longer be vertically integrated (i.e., have backward integration).
5. Final product costs may be heavily influenced by the costs of subcontractors.
6. Luxor may not be able to remain a low cost supplier.
7. Layoffs will be inevitable, but perhaps not in the near term.
8. The marketing and selling of products may need to change. Can Luxor still market products as a low-cost, high quality, state-of-the-art manufacturer?
9. Price-cutting by Luxor's competitors could have a serious impact on Luxor's future ability to survive.

The list presented by marketing demonstrated that there was a serious threat to Luxor's growth and even survival. Engineering then prepared a list of alternative courses of action that would enable Luxor to maintain its technical leadership position:

1. Luxor could hire (away from the competition) more staff personnel with pure and applied R&D skills. This would be a costly effort.
2. Luxor could slowly retrain part of its existing labor force using existing, ex-perienced R&D personnel to conduct the training.
3. Luxor could fund seminars and university courses on general R&D methods, as well as R&D methods for telecommunications projects. These programs were available locally.
4. Luxor could use tuition reimbursement funds to pay for distance learning courses (conducted over the Internet). These were full semester programs.
5. Luxor could outsource technical development.
6. Luxor could purchase or license technology from other firms, including com-petitors. This assumed that competitors would agree to this at a reasonable price.
7. Luxor could develop joint ventures/mergers with other companies which, in turn, would probably require Luxor to disclose much of its proprietary knowl-edge.

With marketing's and engineering's lists before them, Luxor's management had to decide which path would be best for the long term.

Questions

1. Can the impact of one specific risk event, such as a technical risk event, create ad-ditional risks, which may or may not be technical risks? Can risk events be inter-related?

2. Does the list provided by marketing demonstrate the likelihood of a risk event or the impact of a risk event?
3. How does one assign probabilities to the marketing list?
4. The seven items in the list provided by engineering are all ways of mitigating certain risk events. If the company follows these suggestions, is it adopting a risk response mode of avoidance, assumption, reduction, or deflection?
5. Would you side with marketing or engineering? What should Luxor do at this point?

Case 3 _____

Altex Corporation

Background

Following World War II, the United States entered into a Cold War with Russia. To win this Cold War, the United States had to develop sophisticated weapon systems with such destructive power that any aggressor knew that the retaliatory capability of the United States could and would inflict vast destruction.

Hundreds of millions of dollars were committed to ideas concerning technology that had not been developed as yet. Aerospace and defense contractors were growing without bounds, thanks to cost-plus-percentage-of-cost contract awards. Speed and technological capability were judged to be significantly more important than cost. To make matters worse, contracts were often awarded to the second or third most qualified bidder for the sole purpose of maintaining competition and maximizing the total number of defense contractors.

Contract Award

During this period, Altex Corporation was elated when it learned that it had just been awarded the R&D phase of the Advanced Tactical Missile Program (ATMP). The terms of the contract specified that Altex had to submit to the Army, within 60 days after contract award, a formal project plan for the two-year ATMP effort. Contracts at that time did not require that a risk management plan be developed. A meeting was held with the project manager of R&D to assess the risks in the ATMP effort.

PM: I'm in the process of developing the project plan. Should I also develop a risk management plan as part of the project plan?

Sponsor: Absolutely not! Most new weapon systems requirements are established by military personnel who have no sense of reality about what it takes to develop a weapon system based upon technology that doesn't even exist yet. We'll be lucky if we can deliver 60 to 70 percent of the specification imposed upon us.

PM: But that's not what we stated in our proposal. I wasn't brought on board until after we won the award, so I wasn't privileged to know the thought process that went into the proposal. The proposal even went so far as to imply that we might be able to exceed the specification limits, and now you're saying that we should be happy with 60 to 70 percent.

Sponsor: We say what we have to say to win the bid. Everyone does it. It is common practice. Whoever wins the R&D portion of the contract will also be first in line for the manufacturing effort and that's where the megabucks come from! If we can achieve 60 to 70 percent of specifications, it should placate the Army enough to give us a follow-on contract. If we told the Army the true cost of developing the technology to meet the specification limits, we would never get the contract. The program might even be canceled. The military people want this weapon system. They're not stupid! They know what is happening, and they do not want to go to their superiors for more money until later on, downstream, after approval by DoD and project kickoff. The government wants the lowest cost and we want long-term, follow-on production contracts, which can generate huge profits.

PM: Aren't we simply telling lies in our proposal?

Sponsor: My engineers and scientists are highly optimistic and believe they can do the impossible. This is how technological breakthroughs are made. I prefer to call it "over-optimism of technical capability" rather than "telling lies." If my engineers and scientists have to develop a risk management plan, they may become pessimistic, and that's not good for us!

PM: The problem with letting your engineers and scientists be optimistic is that they become reactive rather than proactive thinkers. Without proactive thinkers, we end up with virtually no risk management or contingency plans. When problems surface that require significantly more in the way of resources than we budgeted for, we will be forced to accept crisis management as a way of life. Our costs will increase, and that's not going to make the Army happy.

Sponsor: But the Army won't penalize us for failing to meet cost or for allowing the schedule to slip. If we fail to meet at least 60 to 70 percent of the specification limits, however, then we may well be in trouble. The Army knows there will be a follow-on contract request if we cannot meet specification limits. I consider 60 to 70 percent of the specifications to be the minimum acceptable limits for the Army. The Army wants the program kicked off right now.

Another important point is that long-term contracts and follow-on production contracts allow us to build up a good working relationship with the Army. This is critical. Once we get the initial contract, as we did, the Army will always work with us for follow-on efforts. Whoever gets the R&D effort will almost always get the lucrative production contract. Military officers are under pressure to work with us because their careers may be in jeopardy if they have to tell their superiors that millions of dollars were awarded to the wrong defense contractor. From a career standpoint, the military officers are better off allowing us to downgrade the requirements than admitting that a mistake was made.

PM: I'm just a little nervous managing a project that is so optimistic that major advances in the state of the art must occur to meet specifications. This is why I want to prepare a risk management plan.

Sponsor: You don't need a risk management plan when you know you can spend as much as you want and also let the schedule slip. If you prepare a risk management plan, you will end up exposing a multitude of risks, especially technical risks. The Army might not know about many of these risks, so why expose them and open up Pandora's box? Personally, I believe that the Army does already know many of these risks, but does not want them publicized to their superiors.

If you want to develop a risk management plan, then do it by yourself, and I really mean by yourself. Past experience has shown that our employees will be talking informally to Army personnel at least two to three times a week. I don't want anyone telling the customer that we have a risk management plan. The customer will obviously want to see it, and that's not good for us.

If you are so incensed that you feel obligated to tell the customer what you're doing, then wait about a year and a half. By that time, the Army will have made a considerable investment in both us and the project, and they'll be locked into us for follow-on work. Because of the strategic timing and additional costs, they will never want to qualify a second supplier so late in the game. Just keep the risk management plan to yourself for now.

If it looks like the Army might cancel the program, then we'll show them the risk management plan, and perhaps that will keep the program alive.

Questions

1. Why was a risk management plan considered unnecessary?
2. Should risk management planning be performed in the proposal stage or after contract award, assuming that it must be done?
3. Does the customer have the right to expect the contractor to perform risk analysis and develop a risk management plan if it is not called out as part of the contractual statement of work?
4. Would Altex have been more interested in developing a risk management plan if the project were funded entirely from within?
5. How effective will the risk management plan be if developed by the project manager in seclusion?
6. Should the customer be allowed to participate in or assist the contractor in developing a risk management plan?
7. How might the Army have responded if they were presented with a risk management plan early during the R&D activities?
8. How effective is a risk management plan if cost overruns and schedule slippages are always allowed?
9. How can severe optimism or severe pessimism influence the development of a risk management plan?
10. How does one develop a risk management plan predicated upon needed advances in the state of the art?
11. Can the sudden disclosure of a risk management plan be used as a stopgap measure to prevent termination of a potentially failing project?
12. Can risk management planning be justified on almost all programs and projects?

Case 4

Acme Corporation

Background

Acme Corporation embarked upon an optimistic project to develop a new product for the marketplace. Acme's scientific community made a technical breakthrough and now the project appears to be in the development stage, more than being pure or applied research.

The product is considered to be high tech. If the product can be launched within the next four months, Acme expects to dominate the market for at least a year or so until the competition catches up. Marketing has stated that the product must sell for not more than $150–$160 per unit to be the cost-focused market leader.

Acme uses a project management methodology for all multifunctional projects. The methodology has six life cycle phases:

- Preliminary planning
- Detailed planning
- Execution/design selection
- Prototyping
- Testing/buyoff
- Production

At the end of each life cycle phase a gate/phase review meeting is held with the project sponsor and other appropriate stakeholders. Gate review meetings are formal meetings. The company has demonstrated success following this methodology for managing projects.

At the end of the second life cycle stage of this project, detailed planning, a meeting is held with just the project manager and the project sponsor. The purpose of the meeting is to review the detailed plan and identify any future problem areas that will require involvement by the project sponsor.

The Meeting

Sponsor: I simply do not understand this document you sent me entitled "Risk Management Plan." All I see is a work breakdown structure with work packages at level 5 of the WBS accompanied by almost 100 risk events. Why am I looking at more than 100 risk events? Furthermore, they're not categorized in any manner. Doesn't our project management methodology provide any guidance on how to do this?

PM: All of these risk events can and will impact the design of the final product. We must be sure we select the right design at the lowest risk. Unfortunately, our project management methodology does not include any provisions or guidance on how to develop a risk management plan. Perhaps it should.

Sponsor: I see no reason for an in-depth analysis of 100 or so risk events. That's too many. Where are the probabilities and expected outcomes or damages?

PM: My team will not be assigning probabilities or damages until we get closer to prototype development. Some of these risk events may go away altogether.

Sponsor: Why spend all of this time and money on risk identification if the risks can go away next month? You've spent too much money doing this. If you spend the same amount of money on all of the risk management steps, then we'll be way over budget.

PM: We haven't looked at the other risk management steps yet, but I believe all of the remaining steps will require less than 10 percent of the budget we used for risk identification. We'll stay on budget.

Questions

1. Was the document given to the sponsor a risk management plan?
2. Did the project manager actually perform effective risk management?
3. Was the appropriate amount of time and money spent identifying the risk events?
4. Should one step be allowed to "dominate" the entire risk management process?
5. Are there any significant benefits to the amount of work already done for risk identification?
6. Should the 100 or so risk events identified have been categorized? If so, how?
7. Can probabilities of occurrence and expected outcomes (i.e., damage) be accurately assigned to 100 risk events?
8. Should a project management methodology provide guidance for the development of a risk management plan?
9. Given the life cycle phases in the case study, in which phase would it be appropriate to identify the risk management plan?
10. What are your feelings on the project manager's comments that he must wait until the prototyping phase to assign probabilities and outcomes?

Quantum Telecom

In June of 1998, the executive committee of Quantum Telecom reluctantly approved two R&D projects that required technical breakthroughs. To make matters worse, the two products had to be developed by the summer of 1999 and introduced into the marketplace quickly. The life expectancy of both products was estimated to be less than one year because of the rate of change in technology. Yet, despite these risks, the two projects were fully funded. Two senior executives were assigned as the project sponsors, one for each project.

Quantum Telecom had a world-class project management methodology with five life cycle phases and five gate review meetings. The gate review meetings were go/no-go decision points based upon present performance and future risks. Each sponsor was authorized and empowered to make any and all decisions relative to projects, including termination.

Company politics always played an active role in decisions to terminate a project. Termination of a project often impacted the executive sponsor's advancement opportunities because the projects were promoted by the sponsors and funded through the sponsor's organization.

During the first two gate review meetings, virtually everyone recommended the termination of both projects. Technical breakthroughs seemed unlikely, and the schedule appeared unduely optimistic. But terminating the projects this early would certainly not reflect favorably upon the sponsors. Reluctantly, both sponsors agreed to continue the projects to the third gate in hopes of a "miracle."

During the third gate review, the projects were still in peril. Although the technical breakthrough opportunity now seemed plausible, the launch date would have to be slipped, thus giving Quantum Telecom a window of only six months to sell the products before obsolescence would occur.

By the fourth gate review, the technical breakthrough had not yet occurred but did still seem plausible. Both project managers were still advocating the cancellation of the projects, and the situation was getting worse. Yet, in order to "save face" within the corporation, both sponsors allowed the projects to continue to completion. They asserted that, "If the new products could not be sold in sufficient quantity to recover the R&D costs, then the fault lies with marketing and sales, not with us." The sponsors were now off the hook, so to speak.

Both projects were completed six months late. The salesforce could not sell as much as one unit, and obsolescence occurred quickly. Marketing and sales were blamed for the failures, not the project sponsors.

Questions
1. How do we eliminate politics from gate review meetings?
2. How can we develop a methodology where termination of a project is not viewed as a failure?
3. Were the wrong people assigned as sponsors?
4. What options are available to a project manager when there exists a disagreement between the sponsor and the project manager?
5. Can your answer to the above question be outlined as part of the project management methodology?

Case 6

Lakes Automotive

Lakes Automotive is a Detroit-based tier one supplier to the auto industry. Between 1995 and 1999, Lakes Automotive installed a project management methodology based upon nine life cycle phases. All 60,000 employees world-wide accepted the methodology and used it. Management was pleased with the results. Also, Lakes Automotive's customer base was pleased with the methodology and provided Lakes Automotive with quality award recognition that everyone believed was attributed to how well the project management methodology was executed.

In February 2000, Lakes Automotive decided to offer additional products to their customers. Lakes Automotive bought out another tier one supplier, Pelex Automotive Products (PAP). PAP also had a good project management reputation and also provided quality products. Many of their products were similar to those provided by Lakes Automotive.

Since the employees from both companies would be working together closely, a singular project management methodology would be required that would be acceptable to both companies. PAP had a good methodology based upon five life cycle phases. Both methodologies had advantages and disadvantages, and both were well liked by their customers.

Questions
1. How do companies combine their methodologies?
2. How do you get employees to change work habits that have proven to be successful?
3. What influence should a customer have in redesigning a methodology that has been proven to be successful?
4. What if the customers want the existing methodologies left intact?
5. What if the customers are unhappy with the new combined methodology?

Case 7

Ferris HealthCare, Inc.

In July of 1999, senior management at Ferris recognized that its future growth could very well be determined by how quickly and how well it implemented project management. For the past several years, line managers had been functioning as project managers while still managing their line groups. The projects came out with the short end of the stick, most often late and over budget, because managers focused on line activities rather than project work. Everyone recognized that project management needed to be an established career path position and that some structured process had to be implemented for project management.

A consultant was brought into Ferris to provide initial project management training for 50 out of the 300 employees targeted for eventual project management training. Several of the employees thus trained were then placed on a committee with senior management to design a project management stage-gate model for Ferris.

After two months of meetings, the committee identified the need for three different stage-gate models: one for information systems, one for new products/services provided, and one for bringing on board new corporate clients. There were several similarities among the three models. However, personal interests dictated the need for three methodologies, all based upon rigid policies and procedures.

After a year of using three models, the company recognized it had a problem deciding how to assign the right project manager to the right project. Project managers had to be familiar with all three methodologies. The alternative, considered impractical, was to assign only those project managers familiar with that specific methodology.

After six months of meetings, the company consolidated the three methodologies into a single methodology, focusing more upon guidelines than on policies and procedures. The entire organization appeared to support the new singular methodology. A consultant was brought in to conduct the first three days of a four-day training program for employees not yet trained in project management. The fourth day was taught by internal personnel with a focus on how to use the new methodology. The success to failure ratio on projects improved dramatically.

Questions

1. Why was it so difficult to develop a singular methodology from the start?
2. Why were all three initial methodologies based upon policies and procedures?
3. Why do you believe the organization later was willing to accept a singular methodology?
4. Why was the singular methodology based upon guidelines rather than policies and procedures?
5. Did it make sense to have the fourth day of the training program devoted to the methodology and immediately attached to the end of the three-day program?
6. Why was the consultant not allowed to teach the methodology?

Case 8

Clark Faucet Company*

Background

By 1999, Clark Faucet Company had grown into the third largest supplier of faucets for both commercial and home use. Competition was fierce. Consumers would evaluate faucets on artistic design and quality. Each faucet had to be available in at least 25 different colors. Commercial buyers seemed more interested in the cost than the average consumer, who viewed the faucet as an object of art, irrespective of price.

Clark Faucet Company did not spend a great deal of money advertising on the radio or on television. Some money was allocated for ads in professional journals. Most of Clark's advertising and marketing funds were allocated to the two semiannual home and garden trade shows and the annual builders trade show. One large builder could purchase more than 5,000 components for the furnishing of one newly constructed hotel or one apartment complex. Missing an opportunity to display the new products at these trade shows could easily result in a 6 to 12 month window of lost revenue.

Culture

Clark Faucet had a noncooperative culture. Marketing and engineering would never talk to one another. Engineering wanted the freedom to design new products, whereas marketing wanted final approval to make sure that what was designed could be sold.

The conflict between marketing and engineering became so fierce that early attempts to implement project management failed. Nobody wanted to be the project

---

*Reprinted from H. Kerzner, *Applied Project Management: Best Practices on Implementation.* New York: Wiley, 2000, pp. 369–371.

manager. Functional team members refused to attend team meetings and spent most of their time working on their own "pet" projects rather than doing the required team work. Their line managers also showed little interest in supporting project management.

Project management became so disliked that the procurement manager refused to assign any of his employees to project teams. Instead, he mandated that all project work come through him. He eventually built up a large brick wall around his employees. He claimed that this would protect them from the continuous conflicts between engineering and marketing.

The Executive Decision

The executive council mandated that another attempt to implement good project management practices must occur quickly. Project management would be needed not only for new product development but also for specialty products and enhancements. The vice presidents for marketing and engineering reluctantly agreed to try and patch up their differences, but did not appear confident that any changes would take place.

Strange as it may seem, nobody could identify the initial cause of the conflicts or how the trouble actually began. Senior management hired an external consultant to identify the problems, provide recommendations and alternatives, and act as a mediator. The consultant's process would have to begin with interviews.

Engineering Interviews

The following comments were made during engineering interviews:

- "We are loaded down with work. If marketing would stay out of engineering, we could get our job done."
- "Marketing doesn't understand that there's more work for us to do other than just new product development."
- "Marketing personnel should spend their time at the country club and in bar rooms. This will allow us in engineering to finish our work uninterrupted!"
- "Marketing expects everyone in engineering to stop what they are doing in order to put out marketing fires. I believe that most of the time the problem is that marketing doesn't know what they want up front. This leads to change after change. Why can't we get a good definition at the beginning of each project?"

Marketing Interviews

- "Our livelihood rests on income generated from trade shows. Since new product development is 4–6 months in duration, we have to beat up on engineering to make sure that our marketing schedules are met. Why can't engineering understand the importance of these trade shows?"
- "Because of the time required to develop new products [4–6 months], we sometimes have to rush into projects without having a good definition of what is required. When a customer at a trade show gives us an idea for a new product, we rush to get the project underway for introduction at the next trade

show. We then go back to the customer and ask for more clarification and/or specifications. Sometimes we must work with the customer for months to get the information we need. I know that this is a problem for engineering, but it cannot be helped."

The consultant wrestled with the comments but was still somewhat perplexed. "Why doesn't engineering understand marketing's problems?" pondered the consultant. In a follow-up interview with an engineering manager, the following comment was made:

We are currently working on 375 different projects in engineering, and that includes those that marketing requested. Why can't marketing understand our problems?

Questions

1. What is the critical issue?
2. What can be done about it?
3. Can excellence in project management still be achieved and, if so, how? What steps would you recommend?
4. Given the current noncooperative culture, how long will it take to achieve a good cooperative project management culture, and even excellence?
5. What obstacles exist in getting marketing and engineering to agree to a singular methodology for project management?
6. What might happen if benchmarking studies indicate that either marketing or engineering are at fault?
7. Should a singular methodology for project management have a process for the prioritization of projects or should some committee external to the methodology accomplish this?

Hyten Corporation*

On June 5, 1998, a meeting was held at Hyten Corporation, between Bill Knapp, director of sales, and John Rich, director of engineering. The purpose of the meeting was to discuss the development of a new product for a special customer application. The requirements included a very difficult, tight-time schedule. The key to the success of the project would depend on timely completion of individual tasks by various departments.

Bill Knapp: The Business Development Department was established to provide coordination between departments, but they have not really helped. They just stick their nose in when things are going good and mess everything up. They have been out to see several customers, giving them information and delivery dates that we can't possibly meet.

John Rich: I have several engineers who have MBA degrees and are pushing hard for better positions within engineering or management. They keep saying that formal project management is what we should have at Hyten. The informal approach we use just doesn't work all the time. But I'm not sure that just any type of project management will work in our division.

Knapp: Well, I wonder who Business Development will tap to coordinate this project? It would be better to get the manager from inside the organization instead of hiring someone from outside.

_____

*Reprinted from H. Kerzner, *Applied Project Management: Best Practices on Implementation.* New York: Wiley, 2000, pp. 397–406.

Company Background

Hyten Company was founded in 1982 as a manufacturer of automotive components. During the first Gulf War, the company began manufacturing electronic components for the military. After the war, Hyten continued to prosper.

Hyten became one of the major component suppliers for the Space Program, but did not allow itself to become specialized. When the Space Program declined, Hyten developed other product lines, including energy management, building products, and machine tools, to complement their automotive components and electronics fields.

Hyten has been a leader in the development of new products and processes. Annual sales are in excess of $600 million. The Automotive Components Division is one of Hyten's rapidly expanding business areas (see the organizational chart in Exhibit I).

The Automotive Components Division

The management of both the Automotive Components Division and the Corporation itself is young and involved. Hyten has enjoyed a period of continuous growth over the past 15 years as a result of careful planning and having the right people in the right positions at the right time. This is emphasized by the fact that within five years of joining Hyten, every major manager and division head has been promoted to more responsibility within the corporation. The management staff of the Automotive Components Division has an average age of 40 and no one is over 50. Most of the middle managers have MBA degrees and a few have Ph.D.s. Currently, the Automotive Components Division has three manufacturing plants at various locations throughout the country. Central offices and most of the nonproduction functions are located at the main plant. There has been some effort by past presidents to give each separate plant some minimal level of purchasing, quality, manufacturing, engineering, and personnel functions.

Informal Project Management at Hyten Corporation

The Automotive Components Division of Hyten Corporation has an informal system of project management. It revolves around each department handling their own func-

Exhibit I. Organizational chart of the automotive division, Hyten Corporation.

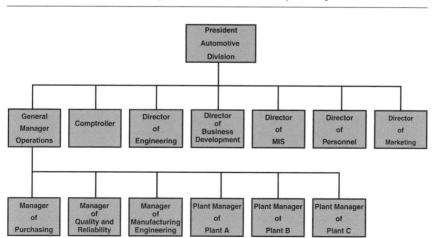

tional area of a given product development or project. Projects have been frequent enough that a sequence of operations has been developed to take a new product from concept to market. Each department knows its responsibilities and what it must contribute to a project.

A manager within the Business Development Department assumes informal project coordination responsibility and calls periodic meetings of the department heads involved. These meetings keep everyone advised of work status, changes to the project, and any problem areas. Budgeting of the project is based on the cost analysis developed after the initial design, while funding is allocated to each functional department based on the degree of its involvement. Funding for the initial design phase is controlled through business development. The customer has very little control over the funding, manpower, or work to be done. The customer, however, dictates when the new product design must be available for integration into the vehicle design, and when the product must be available in production quantities.

The Business Development Department

The Business Development Department, separate from Marketing/Sales, functions as a steering group for deciding which new products or customer requests are to be pursued and which are to be dropped. Factors that they consider in making these decisions are: (1) the company's long- and short-term business plans, (2) current sales forecasts, (3) economic and industry indicators, (4) profit potential, (5) internal capabilities (both volume and technology), and (6) what the customer is willing to pay versus estimated cost.

The duties of Business Development also include the coordination of a project or new product from initial design through market availability. In this capacity, they have no formal authority over either functional managers or functional employees. They act strictly on an informal basis to keep the project moving, give status reports, and report on potential problems. They are also responsible for the selection of the plant that will be used to manufacture the product.

The functions of Business Development were formerly handled as a joint staff function where all the directors would periodically meet to formulate short-range plans and solve problems associated with new products. The department was formally organized three years ago by the then 38-year-old president as a recognition of the need for project management within the Automotive Components Division.

Manpower for the Business Development Department was taken from both outside the company and from within the division. This was done to honor the Corporation's commitment to hire people from the outside only after it was determined that there were no qualified people internally (an area that for years has been a sore spot to the younger managers and engineers).

When the Business Development Department was organized, its level of authority and responsibility was limited. However, the Department's authority and responsibility have subsequently expanded, though at a slow rate. This was done so as not to alienate the functional managers who were concerned that project management would undermine their "empire."

Introduction of Formal Project Management at Hyten Corporation

On July 10, 1998, Wilbur Donley was hired into the Business Development Department to direct new product development efforts. Prior to joining Hyten, he worked as project manager with a company that supplied aircraft hardware to the government. He had worked both as an assistant project manager and as a project manager for five years prior to joining Hyten.

Shortly after his arrival, he convinced upper management to examine the idea of expanding the Business Development group and giving them responsibility for formal project management. An outside consulting firm was hired to give an in-depth seminar on project management to all management and supervisor employees in the Division.

Prior to the seminar, Donley talked to Frank Harrel, manager of quality and reliability, and George Hub, manager of manufacturing engineering, about their problems and what they thought of project management.

Frank Harrel is 37 years old, has an MBA degree, and has been with Hyten for five years. He was hired as an industrial engineer and three years ago was promoted to manager of quality and reliability. George Hub is 45 years old and has been with Hyten for 12 years as manager of manufacturing engineering.

Wilbur Donley: Well, Frank, what do you see as potential problems to the timely completion of projects within the Automotive Components Division?

Frank Harrel: The usual material movement problems we always have. We monitor all incoming materials in samples and production quantities, as well as in-process checking of production and finished goods on a sampling basis. We then move to 100 percent inspection if any discrepancies are found. Marketing and Manufacturing people don't realize how much time is required to inspect for either internal or customer deviations. Our current manpower requires that schedules be juggled to accommodate 100 percent inspection levels on "hot items." We seem to be getting more and more items at the last minute that must be done on overtime.

Donley: What are you suggesting? A coordination of effort with marketing, purchasing, production scheduling, and the manufacturing function to allow your department to perform their routine work and still be able to accommodate a limited amount of high-level work on "hot" jobs?

Harrel: Precisely, but we have no formal contact with these people. More open lines of communication would be of benefit to everyone.

Donley: We are going to introduce a more formal type of project management than has been used in the past so that all departments who are involved will actively participate in the planning cycle of the project. That way we they will remain aware of how they affect the function of other departments and prevent overlapping of work. We should be able to stay on schedule and get better cooperation.

Harrel: Good, I'll be looking forward to the departure from the usual method of handling a new project. Hopefully, it will work much better and result in fewer problems.

Donley: How do you feel, George, about improving the coordination of work among various departments through a formal project manager?

George Hub: Frankly, if it improves communication between departments, I'm all in favor of the change. Under our present system, I am asked to make estimates of cost and lead times to implement a new product. When the project begins, the Product

Design group starts making changes that require new cost figures and lead times. These changes result in cost overruns and in not meeting schedule dates. Typically, these changes continue right up to the production start date. Manufacturing appears to be the bad guy for not meeting the scheduled start date. We need someone to coordinate the work of various departments to prevent this continuous redoing of various jobs. We will at least have a chance at meeting the schedule, reducing cost, and improving the attitude of my people.

Personnel Department's View of Project Management

After the seminar on project management, a discussion was held between Sue Lyons, director of personnel, and Jason Finney, assistant director of personnel. The discussion was about changing the organization structure from informal project management to formal project management.

Sue Lyons: Changing over would not be an easy road. There are several matters to be taken under consideration.

Jason Finney: I think we should stop going to outside sources for competent people to manage new projects that are established within Business Development. There are several competent people at Hyten who have MBA's in Systems/Project Management. With that background and their familiarity with company operations, it would be to the company's advantage if we selected personnel from within our organization.

Lyons: Problems will develop whether we choose someone from inside the company or from an outside source.

Finney: However, if the company continues to hire outsiders into Business Development to head new projects, competent people at Hyten are going to start filtering to places of new employment.

Lyons: You are right about the filtration. Whoever is chosen to be a project manager must have qualifications that will get the job done. He or she should not only know the technical aspect behind the project, but should also be able to work with people and understand their needs. Project managers have to show concern for team members and provide them with work challenge. Project managers must work in a dynamic environment. This often requires the implementation of change. Project managers must be able to live with change and provide necessary leadership to implement the change. It is the project manager's responsibility to develop an atmosphere to allow people to adapt to the changing work environment.

In our department alone, the changes to be made will be very crucial to the happiness of the employees and the success of projects. They must feel they are being given a square deal, especially in the evaluation procedure. Who will do the evaluation? Will the functional manager be solely responsible for the evaluation when, in fact, he or she might never see the functional employee for the duration of a project? A functional manager cannot possibly keep tabs on all the functional employees who are working on different projects.

Finney: Then the functional manager will have to ask the project managers for evaluation information.

Lyons: I can see how that could result in many unwanted situations. To begin with, say the project manager and the functional manager don't see eye to eye on things. Granted, both should be at the same grade level and neither one has authority over the other, but let's say there is a situation where the two of them disagree as to either direction or quality of work. That puts the functional employee in an awkward position. Any employee will have the tendency of bending toward the individual who signs his or her promotion and evaluation form. This can influence the project manager into recommending an evaluation below par regardless of how the functional employee performs. There is also the situation where the employee is on the project for only a couple of weeks, and spends most of his or her time working alone, never getting a chance to know the project manager. The project manager will probably give the functional employee an average rating, even though the employee has done an excellent job. This results from very little contact. Then what do you do when the project manager allows personal feelings to influence his or her evaluation of a functional employee? A project manager who knows the functional employee personally might be tempted to give a strong or weak recommendation, regardless of performance.

Finney: You seem to be aware of many difficulties that project management might bring.

Lyons: Not really, but I've been doing a lot of homework since I attended that seminar on project management. It was a good seminar, and since there is not much written on the topic, I've been making a few phone calls to other colleagues for their opinions on project management.

Finney: What have you learned from these phone calls?

Lyons: That there are more personnel problems involved. What do you do in this situation? The project manager makes an excellent recommendation to the functional manager. The functional employee is aware of the appraisal and feels he or she should be given an above average pay increase to match the excellent job appraisal, but the functional manager fails to do so. One personnel manager from another company incorporating project management ran into problems when the project manager gave an employee of one grade level responsibilities of a higher grade level. The employee did an outstanding job taking on the responsibilities of a higher grade level and expected a large salary increase or a promotion.

Finney: Well, that's fair, isn't it?

Lyons: Yes, it seems fair enough, but that's not what happened. The functional manager gave an average evaluation and argued that the project manager had no business giving the functional employee added responsibility without first checking with him. So, then what you have is a disgruntled employee ready to seek employment elsewhere. Also, there are some functional managers who will only give above-average pay increases to those employees who stay in the functional department and make that manager look good.

Lyons: Right now I can see several changes that would need to take place. The first major change would have to be attitudes toward formal project management and hiring procedures. We do have project management here at Hyten but on an informal basis. If we could administer it formally, I feel we could do the company a great service. If we seek project managers from within, we could save on time and money. I could devote more time and effort on wage and salary grades and job descriptions. We would need to revise our evaluation forms—presently they are not adequate. Maybe we should develop more than one evaluation form: one for the project manager to fill out and give to the functional manager, and a second form to be completed by the functional manager for submission to Personnel.

Finney: That might cause new problems. Should the project manager fill out his or her evaluation during or after project completion?

Lyons: It would have go be after project completion. That way an employee who felt unfairly evaluated would not feel tempted to screw up the project. If an employee felt the work wasn't justly evaluated, that employee might decide not to show up for a few days—these few days of absence could be most crucial for timely project completion.

Finney: How will you handle evaluation of employees who work on several projects at the same time? This could be a problem if employees are really enthusiastic about one project but not about another. They could do a terrific job on the project they are interested in and slack off on other projects. You could also have functional people working on departmental jobs but charging their time to the project overhead. Don't we have exempt and nonexempt people charging to projects?

Lyons: See what I mean? We can't just jump into project management and expect a bed of roses. There will have to be changes. We can't put the cart before the horse.

Finney: I realize that, Sue, but we do have several MBA people working here at Hyten who have been exposed to project management. I think that if we start putting our heads together and take a systematic approach to this matter, we will be able to pull this project together nicely.

Lyons: Well, Jason, I'm glad to see that you are for formal project management. We will have to approach top management on the topic. I would like you to help coordinate an equitable way of evaluating our people and to help develop the appropriate evaluation forms.

Project Management as Seen by the Various Departments

The general manager arranged through the personnel department to interview various managers on a confidential basis. The purpose of the interview was to evaluate the overall acceptance of the concept of formal project management. The answers to the question, "How will project management affect your department?" were as follows:

Frank Harrel, quality and reliability manager:

> Project management is the actual coordination of the resources of functional departments to achieve the time, cost, and performance goals of the project. As a consequence, personnel interfacing is an important component toward the success of the project. In terms of quality control, it means less of the attitude of the structured workplace where quality is viewed as having the function of finding defects and, as a result, is looked upon as a hindrance to production. It means that the attitude toward quality control will change to one of interacting with other departments to minimize manufacturing problems. Project management reduces suboptimization among functional areas and induces cooperation. Both company and department goals can be achieved. It puts an end to the "can't see the forest for the trees" syndrome.

Harold Grimes, plant manager:

I think that formal project management will give us more work than long-term benefits. History indicates that we hire more outside people for new positions than we promote from within. Who will be hired into these new project management jobs? We are experiencing a lot of backlash from people who are required to teach new people the ropes. In my opinion, we should assign inside MBA graduates with project management training to head up projects and not hire an outsider as a formal project manager. Our present system would work fine if inside people were made the new managers in the Business Development Department.

Herman Hall, director of MIS:

I have no objections to the implementation of formal project management in our company. I do not believe, however, that it will be possible to provide the reports needed by this management structure for several years. This is due to the fact that most of my staff are deeply involved in current projects. We are currently working on the installation of minicomputers and on-line terminals throughout the plant. These projects have been delayed by the late arrival of new equipment, employee sabotage, and various start-up problems. As a result of these problems, one group admits to being six months behind schedule and the other group, although on schedule, is 18 months from their scheduled completion date. The rest of the staff currently assigned to maintenance projects consists of two systems analysts who are nearing retirement and two relatively inexperienced programmers. So, as you can readily see, unless we break up the current project teams and let those projects fall further behind schedule, it will be difficult at this time to put together another project team

The second problem is that even if I could put together a staff for the project, it might take up to two years to complete an adequate information system. Problems arise from the fact that it will take time to design a system that will draw data from all the functional areas. This design work will have to be done before the actual programming and testing could be accomplished. Finally, there would be a debugging period when we receive feedback from the user on any flaws in the system or enhancements that might be needed. We could not provide computer support to an "overnight" change to project management.

Bob Gustwell, scheduling manager:

I am happy with the idea of formal project management, but I do see some problems implementing it. Some people around here like the way we do things now. It is a natural reaction for employees to fight against any changes in management style.

But don't worry about the scheduling department. My people will like the change to formal project management. I see this form of management as a way to minimize, of not eliminate, schedule changes. Better planning on the part of both department and project managers will be required, and the priorities will be set at corporate level. You can count on our support because I'm tired of being caught between production and sales.

John Rich, director of engineering:

It seems to me that project management will only mess things up. We now have a good flowing chain of command in our organization. This new matrix will only create problems. The engineering department, being very technical, just can't take direction from anyone outside the department. The project office will start to skimp on specifications just to save time and dollars. Our products are too technical to allow schedules and project costs to affect engineering results.

Bringing in someone from the outside to be the project manager will make things worse. I feel that formal project management should not be implemented at Hyten. Engineering has always directed the projects, and we should keep it that way. We shouldn't change a winning combination.

Fred Kuncl, plant engineering:

I've thought about the trade-offs involved in implementing formal project management at Hyten and feel that plant engineering cannot live with them. Our departmental activities are centered around highly unpredictable circumstances, which sometimes involve rapidly changing priorities related to the production function. We in plant engineering must be able to respond quickly and appropriately to maintenance activities directly related to manufacturing activities. Plant engineering is also responsible for carrying out critical preventive maintenance and plant construction projects.

Project management would hinder our activities because project management responsibilities would burden our manpower with additional tasks. I am against project management because I feel that it is not in the best interest of Hyten. Project management would weaken our department's functional specialization because it would require cross-utilization of resources, manpower, and negotiation for the services critical to plant engineering.

Bill Knapp, director of marketing:

I feel that the seminar on formal project management was a good one. Formal project management could benefit Hyten. Our organization needs to focus in more than one direction at all times. In order to be successful in today's market, we must concentrate on giving all our products sharp focus. Formal project management could be a good way of placing individual emphasis on each of the products of our company. Project management would be especially advantageous to us because of our highly diversified product lines. The organization needs to efficiently allocate resources to projects, products, and markets. We cannot afford to have expensive resources sitting idle. Cross-utilization and the consequent need for negotiation ensures that resources are used efficiently and in the organization's best overall interest.

We can't afford to continue to carry on informal project management in our business. We are so diversified that all of our products can't be treated alike. Each product has different needs. Besides, the nature of a team effort would strengthen our organization.

Stanley Grant, comptroller:

In my opinion, formal project management can be profitably applied in our organization. Management should not, however, expect that project management would gain instant acceptance by the functional managers and functional employees, including the finance department personnel.

The implementation of formal project management in our organization would have an impact on our cost control system and internal control system, as well.

In the area of cost control, project cost control techniques have to be formalized and installed. This would require the accounting staff to: (1) beak comprehensive cost summaries into work packages, (2) prepare commitment reports for "technical decision makers," (3) approximate report data and (4) concentrate talent on major problems and opportunities. In project management, cost commitments on a project are made when various functional departments, such as engineering, manufacturing and marketing, make technical decisions to take some kind of action. Conventional accounting reports do not show the cost effects of these technical decisions until it is too late to reconsider. We would need to provide the project manager with cost commitment reports at each decision state to enable him or her to judge when costs are getting out of control. Only by receiving such timely cost commitment reports, could the project manager take needed corrective actions and be able to approximate the cost effect of each technical decision. Providing all these reports, however, would require additional personnel and expertise in our department.

In addition, I feel that the implementation of formal project management would increase our responsibilities in finance department. We would need to conduct project audits, prepare periodic comparisons of actual versus projected costs and actual versus programmed manpower allocation, update projection reports and funding schedules, and sponsor cost-improvement programs.

In the area of internal control, we will need to review and modify our existing internal control system to effectively meet our organization's goals related to project management. A careful and proper study and evaluation of existing internal control procedures should be conducted to determine the extent of the tests to which our internal auditing procedures are to be restricted. A thorough understanding of each project we undertake must be required at all times.

I'm all in favor of formal project management, provided management would allocate more resources to our department so we could maintain the personnel necessary to perform the added duties, responsibilities, and expertise required.

After the interviews, Sue Lyons talked to Wilbur Donley about the possibility of adopting formal project management. As she put it,

You realize that regardless of how much support there is for formal project management, the general manager will probably not allow us to implement it for fear it will affect the performance of the Automotive Components Division.

Questions

1. What are some of the major problems facing the management of Hyten in accepting formalized project management? (Include attitude problems/personality problems.)
2. Do any of the managers appear to have valid arguments for their beliefs as to why formal project management should not be considered?
3. Are there any good reasons why Hyten should go to formal project management?
4. Has Hyten taken a reasonable approach toward implementing formal project management?
5. Has Hyten done anything wrong?
6. Should formal project management give employees more room for personal growth?
7. Will formalized project management make it appear as though business development has taken power away from other groups?
8. Were the MBAs exposed to project management?
9. Were the organizational personnel focusing more on the problems (disadvantages) or advantages of project management?
10. What basic fears do employees have in considering organizational change to formal project management?
11. Must management be sold on project management prior to implementation?
12. Is it possible that some of the support groups cannot give immediate attention to such an organizational change?
13. Do functional managers risk a loss of employee loyalty with the new change?
14. What recommendations would you make to Hyten Corporation?
15. Is it easier or more difficult to implement a singular methodology for project management after the company has adopted formal project management rather than informal project management?
16. Is strategic planning for project management easier or more difficult to perform with formal project management in place?

Como Tool and Die (A)*

Como Tool and Die was a second-tier component supplier to the auto industry. Their largest customer was Ford Motor Company. Como had a reputation for delivering a quality product. During the 1980s and the early 1990s, Como's business grew because of their commitment to quality. Emphasis was on manufacturing operations, and few attempts were made to use project management. All work was controlled by line managers who, more often than not, were overburdened with work.

The culture at Como underwent a rude awakening in 1996. In the summer of 1996, Ford Motor Company established four product development objectives for both tier one and tier two suppliers:

- Lead time: 25–35 percent reduction
- Internal resources: 30–40 percent reduction
- Prototypes: 30–35 percent reduction (time and cost)
- Continuous process improvement and cost reductions

The objectives were aimed at consolidation of the supply base with larger commitments to tier one suppliers, who would now have greater responsibility in vehicle development, launch, process improvement, and cost reduction. Ford had established a time frame of 24 months for achievement of the objectives. The ultimate goal for Ford would be the creation of one global, decentralized vehicle development system that would benefit from the efficiency and technical capabilities of the original equipment manufacturers (OEMs) and the subsupplier infrastructure.

---

*Fictitious case. Reprinted from H. Kerzner, *Applied Project Management: Best Practices on Implementation.* New York: Wiley, 2000, pp. 420–423.

Strategic Redirection: 1996

Como realized that it could no longer compete on quality alone. The marketplace had changed. The strategic plan for Como was now based upon maintaining an industry leadership position well into the twenty-first century. The four basic elements of the strategic plan included:

- First to market (faster development and tooling of the right products)
- Flexible processes (quickly adaptable to model changes)
- Flexible products (multiple niche products from shared platforms and a quick-to-change methodology)
- Lean manufacturing (low cost, high quality, speed, and global economies of scale)

The implementation of the strategy mandated superior project management performance, but changing a 60-year culture to support project management would not be an easy task.

The president of the company established a task force to identify the cultural issues of converting over to an informal project management system. The president believed that project management would eventually become the culture and, therefore, that the cultural issues must be addressed first. The following list of cultural issues was identified by the task force:

- Existing technical, functional departments currently do not adequately support the systemic nature of projects as departmental and individual objectives are not consistent with those of the project and the customer.
- Senior management must acknowledge the movement away from traditional, "over the fence," management and openly endorse the significance of project management, teamwork, and delegation of authority as the future.
- The company must establish a system of project sponsorship to support project managers by trusting them with the responsibility and then empowering them to be successful.
- The company must educate managers in project and risk management and the cultural changes of cross-functional project support; it is in the manager's self interest to support the project manager by providing necessary resources and negotiating for adequate time to complete the work.
- The company must enhance information systems to provide cost and schedule performance information for decision-making and problem resolution.
- Existing informal culture can be maintained while utilizing project management to monitor progress and review costs. Bureaucracy, red tape, and lost time must be eliminated through project management's enhanced communications, standard practices, and goal congruence.

The task force, as a whole, supported the idea of informal project management and believed that all of the cultural issues could be overcome. The task force identified four critical risks and the method of resolution:

- Trusting others and the system.
  - *Resolution:* Training in the process of project management and understanding of the benefits. Interpersonal training to learn to trust in each other and in keeping commitments will begin the cultural change.

- Transform 60 years of tradition in vertical reporting into horizontal project management.
  - *Resolution:* Senior management sponsor the implementation program, participate in training, and fully support efforts to implement project management across functional lines with encouragement and patience as new organizational relationships are forged.
- Capacity constraints and competition for resources.
  - *Resolution:* Work with managers to understand constraints and to develop alternative plans for success. Develop alternative external capacity to support projects.
- Inconsistency in application after introduction.
  - *Resolution:* Set the clear expectation that project management is the operational culture and the responsibility of each manager. Set the implementation of project management as a key measurable for management incentive plans. Establish a model project and recognize the efforts and successes as they occur.

The president realized that project management and strategic planning were related. The president wondered what would happen if the business base would grow as anticipated. Could project management excellence enhance the business base even further? To answer this question, the president prepared a list of competitive advantages that could be achieved through superior project management performance:

- Project management techniques and skills must be enhanced, especially for the larger, complex projects.
- Development of broader component and tooling supply bases would provide for additional capacity.
- Enhanced profitability would be possible through economies of scale to utilize project managers and skilled trades resources more efficiently through balanced workloads and level production.
- Greater purchasing leverage would be possible through larger purchasing volume and sourcing opportunities.
- Disciplined coordination, reporting of project status and proactive project management problem-solving must exist to meet timing schedules, budgets, and customer expectations.
- Effective project management of multitiered supply base will support sales growth beyond existing, capital intensive, internal tooling, and production capacities.

The wheels were set in motion. The president and his senior staff met with all of the employees of Como Tool and Die to discuss the implementation of project management. The president made it clear that he wanted a mature project management system in place within 36 months.

Questions

1. Does Como have a choice in whether or not to accept project management as a culture?
2. How much influence should a customer be able to exert on how the contractors manage projects?

3. Was Como correct in attacking the cultural issues first?
4. Does the time frame of 36 months seem practical?
5. What chance of success do you give Como?
6. What dangers exist when your customers are more knowledgeable than you are concerning project management?
7. Is it possible for your customers' knowledge of project management to influence the way that your organization performs strategic planning for project management?
8. Should your customer, especially if a powerful customer, have an input in the way that your organization performs strategic planning for project management? If so, what type of input should the customer have and on what subject matter?

Case 11 _____
Como Tool and Die (B)*

By 1997, Como had achieved partial success in implementing project management. Lead times were reduced by 10 percent rather than the target of 25–35 percent. Internal resources were reduced by only 5 percent. The reduction in prototype time and cost was 15 percent rather than the expected 30–35 percent.

Como's automotive customers were not pleased with the slow progress and relatively immature performance of Como's project management system. Change was taking place, but not fast enough to placate the customers. Como was on target according to its 36 month schedule to achieve some degree of excellence in project management, but would its customers be willing to wait another two years for completion, or should Como try to accelerate the schedule?

Ford Introduces "Chunk" Management

In the summer of 1997, Ford announced to its suppliers that it was establishing a "chunk" management system. All new vehicle metal structures would be divided into three or four major portions with each chosen supplier (i.e., chunk manager) responsible for all components within that portion of the vehicle. To reduce lead time at Ford and to gain supplier commitment, Ford announced that advanced placement of new work (i.e., chunk managers) would take place without competitive bidding. Target agreements on piece price, tooling cost, and lead time would be established and equitably negotiated later with value engineering work acknowledged.

Chunk managers would be selected based upon superior project management ca-

*Fictitious case. Reprinted from H. Kerzner, *Applied Project Management: Best Practices on Implementation.* New York: Wiley, 2000, pp. 424–425.

pability, including program management skills, coordination responsibility, design feasibility, prototypes, tooling, testing, process sampling, and start of production for components and subassemblies. Chunk managers would function as the second tier component suppliers and coordinate vehicle build for multiple, different vehicle projects at varied stages in the development–tool–launch process.

Strategic Redirection: 1997

Ford Motor Company stated that the selection of the chunk managers would not take place for another year. Unfortunately, Como's plan to achieve excellence would not have been completed by then and its chances to be awarded a chunk management slot were slim.

The automotive division of Como was now at a critical junction. Como's management believed that the company could survive as a low-level supplier of parts, but its growth potential would be questionable. Chunk managers might find it cost-effective to become vertically integrated and produce for themselves the same components that Como manufactured. This could have devastating results for Como. This alternative was unacceptable.

The second alternative required that Como make it clear to Ford Motor Company that Como wished to be considered for a chunk manager contract. If Como were to be selected, then Como's project management systems would have to:

- Provide greater coordination activities than previously anticipated
- Integrate concurrent engineering practices into the company's existing methodology for project management
- Decentralize the organization so as to enhance the working relationship with the customers
- Plan for better resource allocation so as to achieve a higher level of efficiency
- Force proactive planning and decision-making
- Drive out waste and lower cost while improving on-time delivery

There were also serious risks if Como were to become a chunk manager. The company would be under substantially more pressure to meet cost and delivery targets. Most of its resources would have to be committed to complex coordination activities rather than new product development. Therefore, value-added activities for its customers would be diminished. Finally, if Como failed to live up to its customers' expectations as a chunk manager, it might end up losing all automotive work.

The decision was made to inform Ford of Como's interest in chunk management. Now Como realized that its original three-year plan for excellence in project management would have to be completed in 18 months. The question on everyone's mind was: "How?"

Questions

1. What was the driving force for excellence before the announcement of chunk management, and what is it now?
2. How can Como accelerate the learning process to achieve excellence in project management? What steps should management take based upon their learning so far?

3. What are their chances for success? Justify your answer.
4. Should Como compete to become a chunk manager?
5. Can the decision to become a chunk supplier change the way Como performs strategic planning for project management?
6. Can the decision to become a chunk supplier cause an immediate change in Como's singular methodology for project management?
7. If a singular methodology for project management already exists, then how difficult will it be to make major changes to the methodology and what type of resistance, if any, should management expect?

Case 12

Macon, Inc.*

Macon was a 50-year-old company in the business of developing test equipment for the tire industry. The company had a history of segregated departments with very focused functional line managers. The company had two major technical departments: mechanical engineering and electrical engineering. Both departments reported to a vice president for engineering, whose background was always mechanical engineering. For this reason, the company focused all projects from a mechanical engineering perspective. The significance of the test equipment's electrical control system was often minimized when, in reality, the electrical control systems were what made Macon's equipment outperform that of the competition.

Because of the strong autonomy of the departments, internal competition existed. Line managers were frequently competing with one another rather than focusing on the best interest of Macon. Each would hope the other would be the cause for project delays instead of working together to avoid project delays altogether. Once dates slipped, fingers were pointed and the problem would worsen over time.

One of Macon's customers had a service department that always blamed engineering for all of their problems. If the machine was not assembled correctly, it was engineering's fault for not documenting it clearly enough. If a component failed, it was engineering's fault for not designing it correctly. No matter what problem occurred in the field, customer service would always put the blame on engineering.

As might be expected, engineering would blame most problems on production, claiming that production did not assemble the equipment correctly and did not maintain the proper level of quality. Engineering would design a product and then throw it

*Reprinted from H. Kerzner, *Applied Project Management: Best Practices on Implementation.* New York: Wiley, 2000, pp. 431–432.

over the fence to production without ever going down to the manufacturing floor to help with its assembly. Errors or suggestions reported from production to engineering were being ignored. Engineers often perceived the assemblers as incapable of improving the design.

Production ultimately assembled the product and shipped it out to the customer. Oftentimes during assembly the production people would change the design as they saw fit without involving engineering. This would cause severe problems with documentation. Customer service would later inform engineering that the documentation was incorrect, once again causing conflict among all departments.

The president of Macon was a strong believer in project management. Unfortunately, his preaching fell upon deaf ears. The culture was just too strong. Projects were failing miserably. Some failures were attributed to the lack of sponsorship or commitment from line managers. One project failed as the result of a project leader who failed to control scope. Each day the project would fall further behind because work was being added with very little regard for the project's completion date. Project estimates were based upon a "gut feel" rather than upon sound quantitative data.

The delay in shipping dates was creating more and more frustration for the customers. The customers began assigning their own project managers as "watchdogs" to look out for their companies' best interests. The primary function of these "watchdog" project managers was to ensure that the equipment purchased would be delivered on time and complete. This involvement by the customers was becoming more prominent than ever before.

The president decided that action was needed to achieve some degree of excellence in project management. The question was what action to take, and when.

Questions

1. Where will the greatest resistance for excellence in project management come from?
2. What plan should be developed for achieving excellence in project management?
3. How long will it take to achieve some degree of excellence?
4. Explain the potential risks to Macon if the customer's experience with project management increases while Macon's knowledge remains stagnant.

Case 13 —————————————————————————
The Trophy Project*

The ill-fated Trophy Project as in trouble right from the start. Reichart, who had been an assistant project manager, was involved with the project from its conception. When the Trophy Project was accepted by the company, Reichart was assigned as the project manager. The program schedules started to slip from day one, and expenditures were excessive. Reichart found that the functional managers were charging direct labor time to his project but working on their own "pet" projects. When Reichart complained of this, he was told not to meddle in the functional manager's allocation of resources and budgeted expenditures. After approximately six months, Reichart was requested to make a progress report directly to corporate and division staffs.

Reichart took this opportunity to bare his soul. The report substantiated that the project was forecasted to be one complete year behind schedule. Reichart's staff, as supplied by the line managers, was inadequate to stay at the required pace, let alone make up any time that had already been lost. The estimated cost at completion at this interval showed a cost overrun of at least 20 percent. This was Reichart's first opportunity to tell his story to people who were in a position to correct the situation. The result of Reichart's frank, candid evaluation of the Trophy Project was very predictable. Nonbelievers finally saw the light, and the line managers realized that they had a role to play in the completion of the project. Most of the problems were now out in the open and could be corrected by providing adequate staffing and resources. Corporate staff ordered immediate remedial action and staff support to provide Reichart a chance to bail out his program.

---

*Reprinted from H. Kerzner, *Project Management: A Systems Approach to Planning, Scheduling and Controlling,* 6th ed. New York: Wiley, 1998, pp. 298–300.

The results were not at all what Reichart had expected. He no longer reported to the project office; he now reported directly to the operations manager. Corporate staff's interest in the project became very intense, requiring a 7:00 A.M. meeting every Monday morning for complete review of the project status and plans for recovery. Reichart found himself spending more time preparing paperwork, reports, and projections for his Monday morning meetings than he did administering the Trophy Project. The main concern of corporate was to get the project back on schedule. Reichart spent many hours preparing the recovery plan and establishing manpower requirements to bring the program back onto the original schedule.

Group staff, in order to closely track the progress of the Trophy Project, assigned an assistant program manager. The assistant program manager determined that a sure cure for the Trophy Project would be to computerize the various problems and track the progress through a very complex computer program. Corporate provided Reichart with 12 additional staff members to work on the computer program. In the meantime, nothing changed. The functional managers still did not provide adequate staff for recovery, assuming that the additional manpower Reichart had received from corporate would accomplish that task.

After approximately $50,000 was spent on the computer program to track the problems, it was found that the program objectives could not be handled by the computer. Reichart discussed this problem with a computer supplier and found that $15,000 more was required for programming and additional storage capacity. It would take two months for installation of the additional storage capacity and the completion of the programming. At this point, the decision was made to abandon the computer program.

Reichart was now a year and a half into the program with no prototype units completed. The program was still nine months behind schedule with the overrun projected at 40 percent of budget. The customer had been receiving his reports on a timely basis and was well aware of the fact that the Trophy Project was behind schedule. Reichart had spent a great deal of time with the customer explaining the problems and the plan for recovery. Another problem that Reichart had to contend with was that the vendors who were supplying components for the project were also running behind schedule.

One Sunday morning, while Reichart was in his office putting together a report for the client, a corporate vice president came into his office. "Reichart," he said, "in any project I look at the top sheet of paper and the man whose name appears at the top of the sheet is the one I hold responsible. For this project your name appears at the top of the sheet. If you cannot bail this thing out, you are in serious trouble in this corporation." Reichart did not know which way to turn or what to say. He had no control over the functional managers who were creating the problems, but he was the person who was being held responsible.

After another three months the customer, becoming impatient, realized that the Trophy Project was in serious trouble and requested that the division general manager and his entire staff visit the customer's plant to give a progress and "get well" report within a week. The division general manager called Reichart into his office and said, "Reichart, go visit our customer. Take three or four functional line people with you and try to placate him with whatever you feel is necessary." Reichart and four functional line people visited the customer and gave a four-and-a-half-hour presentation defining the problems and the progress to that point. The customer was very polite and even commented that it was an excellent presentation, but the content was totally unacceptable. The program was still six to eight months late, and the customer de-

manded progress reports on a weekly basis. The customer made arrangements to assign a representative in Reichart's department to be "on-site" at the project on a daily
basis and to interface with Reichart and his staff as required. After this turn of events,
the program became very hectic.

The customer representative demanded constant updates and problem identification and then became involved in attempting to solve these problems. This involvement created many changes in the program and the product in order to eliminate some
of the problems. Reichart had trouble with the customer and did not agree with the
changes in the program. He expressed his disagreement vocally when, in many cases,
the customer felt the changes were at no cost. This caused a deterioration of the relationship between client and producer.

One morning Reichart was called into the division general manager's office and
introduced to Mr. "Red" Baron. Reichart was told to turn over the reins of the Trophy
Project to Red immediately. "Reichart, you will be temporarily reassigned to some
other division within the corporation. I suggest you start looking outside the company
for another job." Reichart looked at Red and asked, "Who did this? Who shot me
down?"

Red was program manager on the Trophy Project for approximately six months,
after which, by mutual agreement, he was replaced by a third project manager. The
customer reassigned his local program manager to another project. With the new team
the Trophy Project was finally completed one year behind schedule and at a 40 percent cost overrun.

Questions

1. Did the project appear to be planned correctly?
2. Did functional management seem to be committed to the project?
3. Did senior management appear supportive and committed?
4. Can a singular methodology for project management be designed to "force" cooperation to occur between groups?
5. Is it possible or even desirable for strategic planning for project management to include ways to improve cooperation and working relationships, or is this beyond the
   scope of strategic planning for project management?

Case 14

The Blue Spider Project*

"This is impossible! Just totally impossible! Ten months ago I was sitting on top of the world. Upper-level management considered me one of the best, if not the best, engineer in the plant. Now look at me! I have bags under my eyes, I haven't slept soundly in the last six months, and here I am, cleaning out my desk. I'm sure glad they gave me back my old job in engineering. I guess I could have saved myself a lot of grief and aggravation had I not accepted the promotion to project manager."

History

Gary Anderson had accepted a position with Parks Corporation right out of college. With a Ph.D. in mechanical engineering, Gary was ready to solve the world's most traumatic problems. At first, Parks Corporation offered Gary little opportunity to do the pure research that he eagerly wanted to undertake. However, things soon changed. Parks grew into a major electronics and structural design corporation during the big boom of the late 1950s and early 1960s when Department of Defense (DoD) contracts were plentiful.

Parks Corporation grew from a handful of engineers to a major DoD contractor, employing some 6,500 people. During the recession of the late 1960s, money became scarce and major layoffs resulted in lowering the employment level to 2,200 employees. At that time, Parks decided to get out of the R&D business and compete as a low-cost production facility while maintaining an engineering organization solely to support production requirements.

After attempts at virtually every project management organizational structure,

---

*Reprinted from H. Kerzner, *Project Management: A Systems Approach to Planning, Scheduling and Controlling,* 6th ed. New York: Wiley, 1998, pp. 494–505.

Parks Corporation selected the matrix form. Each project had a program manager who reported to the director of program management. Each project also maintained an assistant project manager—normally a project engineer—who reported directly to the project manager and indirectly to the director of engineering. The program manager spent most of his time worrying about cost and time, whereas the assistant program manager worried more about technical performance.

With the poor job market for engineers, Gary and his colleagues began taking coursework toward MBA degrees in case the job market deteriorated further.

In 1975, with the upturn in DoD spending, Parks had to change its corporate strategy. Parks had spent the last seven years bidding on the production phase of large programs. Now, however, with the new evaluation criteria set forth for contract awards, those companies winning the R&D and qualification phases had a definite edge on being awarded the production contract. The production contract was where the big profits could be found. In keeping with this new strategy, Parks began to beef up its R&D engineering staff. By 1978, Parks had increased in size to 2,700 employees. The increase was mostly in engineering. Experienced R&D personnel were difficult to find for the salaries that Parks was offering. Parks was, however, able to lure some employees away from the competitors, but relied mostly upon the younger, inexperienced engineers fresh out of college.

With the adoption of this corporate strategy, Parks Corporation administered a new wage and salary program that included job upgrading. Gary was promoted to senior scientist, responsible for all R&D activities performed in the mechanical engineering department. Gary had distinguished himself as an outstanding production engineer during the past several years, and management felt that his contribution could be extended to R&D as well.

In January 1978, Parks Corporation decided to compete for Phase I of the Blue Spider Project, an R&D effort that, if successful, could lead into a $500 million program spread out over 20 years. The Blue Spider Project was an attempt to improve the structural capabilities of the Spartan missile, a short-range tactical missile used by the Army. The Spartan missile was exhibiting fatigue failure after six years in the field. This was three years less than what the original design specifications called for. The Army wanted new materials that could result in a longer life for the Spartan missile.

Lord Industries was the prime contractor for the Army's Spartan Program. Parks Corporation would be a subcontractor to Lord if they could successfully bid and win the project. The criteria for subcontractor selection were based not only on low bid, but also on technical expertise as well as management performance on other projects. Park's management felt that they had a distinct advantage over most of the other competitors because they had successfully worked on other projects for Lord Industries.

The Blue Spider Project Kickoff

On November 3, 1977, Henry Gable, the director of engineering, called Gary Anderson into his office.

Henry Gable: Gary, I've just been notified through the grapevine that Lord will be issuing the RFP for the Blue Spider Project by the end of this month, with a 30-day response period. I've been waiting a long time for a project like this to come along so that I can experiment with some new ideas that I have. This project is going to be my baby all the way! I want you to head up the proposal team. I think it must be an engineer. I'll make sure that you get a good proposal manager to help you. If we start working now, we can get close to two months of research in before proposal submittal. That will give us a one-month's edge on our competitors.

Gary was pleased to be involved in such an effort. He had absolutely no trouble in getting functional support for the R&D effort necessary to put together a technical proposal. All of the functional managers continually remarked to Gary, "This must be a biggy. The director of engineering has thrown all of his support behind you."

On December 2, the RFP was received. The only trouble area that Gary could see was that the technical specifications stated that all components must be able to operate normally and successfully through a temperature range of −65 °F to 145 °F. Current testing indicated the Parks Corporation's design would not function above 130 °F. An intensive R&D effort was conducted over the next three weeks. Everywhere Gary looked, it appeared that the entire organization was working on his technical proposal.

A week before the final proposal was to be submitted, Gary and Henry Gable met to develop a company position concerning the inability of the preliminary design material to be operated above 130 °F.

Gary Anderson: Henry, I don't think it is going to be possible to meet specification requirements unless we change our design material or incorporate new materials. Everything I've tried indicates we're in trouble.

Gable: We're in trouble only if the customer knows about it. Let the proposal state that we expect our design to be operative up to 155 °F. That'll please the customer.

Anderson: That seems unethical to me. Why don't we just tell them the truth?

Gable: The truth doesn't always win proposals. I picked you to head up this effort because I thought that you'd understand. I could have just as easily selected one of our many moral project managers. I'm considering you for program manager after we win the program. If you're going to pull this conscientious crap on me like the other project managers do, I'll find someone else. Look at it this way; later we can convince the customer to change the specifications. After all, we'll be so far downstream that he'll have no choice.

After two solid months of 16-hour days for Gary, the proposal was submitted. On February 10, 1978, Lord Industries announced that Parks Corporation would be awarded the Blue Spider Project. The contract called for a ten-month effort, negotiated at $2.2 million at a firm-fixed price.À

Selecting the Project Manager
Following contract award, Henry Gable called Gary in for a conference.

Gable: Congratulations, Gary! You did a fine job. The Blue Spider Project has great potential for ongoing business over the next ten years, provided that we perform well during the R&D phase. Obviously you're the most qualified person in the plant to head up the project. How would you feel about a transfer to program management?

Anderson: I think it would be a real challenge. I could make maximum use of the MBA degree I earned last year. I've always wanted to be in program management.

Gable: Having several masters' degrees, or even doctorates for that matter, does not guarantee that you'll be a successful project manager. There are three requirements for effective program management: You must be able to communicate both in writing and orally; you must know how to motivate people; and you must be willing to give up your car pool. The last one is extremely important in that program managers must be totally committed and dedicated to the program, regardless of how much time is involved.

But this is not the reason why I asked you to come here. Going from project engineer to program management is a big step. There are only two places you can go from program management—up the organization or out the door. I know of very, very few engineers who failed in program management and were permitted to return.

Anderson: Why is that? If I'm considered to be the best engineer in the plant, why can't I return to engineering?

Gable: Program management is a world of its own. It has its own formal and informal organizational ties. Program managers are outsiders. You'll find out. You might not be able to keep the strong personal ties you now have with your fellow employees. You'll have to force even your best friends to comply with your standards. Program managers can go from program to program, but functional departments remain intact.

I'm telling you all this for a reason. We've worked well together the past several years. But if I sign the release so that you can work for Grey in program management, you'll be on your own, like hiring into a new company. I've already signed the release. You still have some time to think about it.

Anderson: One thing I don't understand. With all of the good program managers we have here, why am I being given this opportunity?

Gable: Almost all of our program managers are over forty-five years old. This resulted from our massive layoffs several years ago when we were forced to lay off the younger, inexperienced program managers. You were selected because of your age and because all of our other program managers have worked only on production-type programs. We need someone at the reins who knows R&D. Your counterpart at Lord Industries will be an R&D type. You have to fight fire with fire.

I have an ulterior reason for wanting you to accept this position. Because of the division of authority between program management and project engineering, I need someone in program management whom I can communicate with concerning R&D work. The program managers we have now are interested only in time and cost. We need a manager who will bend over backwards to get performance also. I think you're that man. You know the commitment we made to Lord when we submitted that proposal. You have to try to achieve that. Remember, this program is my baby. You'll get all the support you need. I'm tied up on another project now. But when it's over, I'll be following your work like a hawk. We'll have to get together occasionally and discuss new techniques.

Take a day or two to think it over. If you want the position, make an appointment to see Elliot Grey, the director of program management. He'll give you the same speech I did. I'll assign Paul Evans to you as chief project engineer. He's a seasoned veteran and you should have no trouble working with him. He'll give you good advice. He's a good man.

The Work Begins

Gary accepted the new challenge. His first major hurdle occurred in staffing the project. The top priority given to him to bid the program did not follow through for staffing. The survival of Parks Corporation depended on the profits received from the production programs. In keeping with this philosophy, Gary found that engineering managers (even his former boss) were reluctant to give up their key people to the Blue Spider Program. However, with a little support from Henry Gable, Gary formed an adequate staff for the program.

Right from the start Gary was worried that the test matrix called out in the technical volume of the proposal would not produce results that could satisfy specifications. Gary had 90 days after go-ahead during which to identify the raw materials that could satisfy specification requirements. Gary and Paul Evans held a meeting to map out their strategy for the first few months.

Anderson: Well, Paul, we're starting out with our backs against the wall on this one. Any recommendations?

Paul Evans: I also have my doubts about the validity of this test matrix. Fortunately, I've been through this before. Gable thinks this is his project and he'll sure as hell try to manipulate us. I have to report to him every morning at 7:30 A.M. with the raw data results of the previous day's testing. He wants to see it before you do. He also stated that he wants to meet with me alone.

Lord will be the big problem. If the test matrix proves to be a failure, we're going to have to change the scope of effort. Remember, this is an FFP contract. If we change the scope of work and do additional work in the earlier phases of the program, then we should prepare a trade-off analysis to see what we can delete downstream so as to not overrun the budget.

Anderson: I'm going to let the other project office personnel handle the administrative work. You and I are going to live in the research labs until we get some results. We'll let the other project office personnel run the weekly team meetings.

For the next three weeks Gary and Paul spent virtually 12 hours per day, 7 days a week, in the research and development lab. None of the results showed any promise. Gary kept trying to set up a meeting with Henry Gable but always found him unavailable.

During the fourth week, Gary, Paul, and the key functional department managers met to develop an alternate test matrix. The new test matrix looked good. Gary and his team worked frantically to develop a new, workable schedule that would not have impact on the second milestone, which was to occur at the end of 180 days. The second milestone was the final acceptance of the raw materials and preparation of production runs of the raw materials to verify that there would be no scale-up differences between lab development and full-scale production.

Gary personally prepared all of the technical handouts for the interchange meeting. After all, he would be the one presenting all of the data. The technical interchange meeting was scheduled for two days. On the first day, Gary presented all of the data, including test results, and the new test matrix. The customer appeared displeased with the progress to date and decided to have its own in-house caucus that evening to go over the material that was presented.

The following morning the customer stated its position: "First of all, Gary, we're quite pleased to have a project manager who has such a command of technology. That's good. But every time we've tried to contact you last month, you were unavailable or had to be paged in the research laboratories. You did an acceptable job presenting the technical data, but the administrative data was presented by your project office personnel. We, at Lord, do not think that you're maintaining the proper balance between your technical and administrative responsibilities. We prefer that you personally give the administrative data and your chief project engineer present the technical data.

"We did not receive any agenda. Our people like to know what will be discussed, and when. We also want a copy of all handouts to be presented at least three days in advance. We need time to scrutinize the data. You can't expect us to walk in here blind and make decisions after seeing the data for ten minutes.

"To be frank, we feel that the data to date is totally unacceptable. If the data does not improve, we will have no choice but to issue a work stoppage order and look for a new contractor. The new test matrix looks good, especially since this is a firm-fixed-price contract. Your company will bear the burden of all costs for the additional work. A trade-off with later work may be possible, but this will depend on the results presented at the second design review meeting, 90 days from now.

"We have decided to establish a customer office at Parks to follow your work more closely. Our people feel that monthly meetings are insufficient during R&D activities. We would like our customer representative to have daily verbal meetings with you or your staff. He will then keep us posted. Obviously, we had expected to review much more experimental data than you have given us.

"Many of our top-quality engineers would like to talk directly to your engineering community, without having to continually waste time by having to go through the project office. We must insist on this last point. Remember, your effort may be only $2.2 million, but our total package is $100 million. We have a lot more at stake than you people do. Our engineers do not like to get information that has been filtered by the project office. They want to help you.

"And last, don't forget that you people have a contractual requirement to prepare complete minutes for all interchange meetings. Send us the original for signature before going to publication."

Although Gary was unhappy with the first team meeting, especially with the requests made by Lord Industries, he felt that they had sufficient justification for their comments. Following the team meeting, Gary personally prepared the complete minutes. "This is absurd," thought Gary. "I've wasted almost one entire week doing nothing more than administrative paperwork. Why do we need such detailed minutes? Can't a rough summary suffice? Why is it that customers want everything documented? That's like an indication of fear. We've been completely cooperative with them. There has been no hostility between us. If we've gotten this much paperwork to do now, I hate to imagine what it will be like if we get into trouble."

A New Role

Gary completed and distributed the minutes to the customer as well as to all key team members.

For the next five weeks testing went according to plan, or at least Gary thought that it had. The results were still poor. Gary was so caught up in administrative paperwork that he hadn't found time to visit the research labs in over a month. On a Wednesday morning, Gary entered the lab to observe the morning testing. Upon arriving in the lab, Gary found Paul Evans, Henry Gable, and two technicians testing a new material, JXB-3.

Gable: Gary, your problems will soon be over. This new material, JXB-3, will permit you to satisfy specification requirements. Paul and I have been testing it for two weeks. We wanted to let you know, but were afraid that if the word leaked out to the customer that we were spending their money for testing materials that were not called out in the program plan, they would probably go crazy and might cancel the contract. Look at these results. They're super!

Anderson: Am I supposed to be the one to tell the customer now? This could cause a big wave.

Gable: There won't be any wave. Just tell them that we did it with our own IR&D funds. That'll please them because they'll think we're spending our own money to support their program.

Before presenting the information to Lord, Gary called a team meeting to present the new data to the project personnel. At the team meeting, one functional manager spoke out: "This is a hell of a way to run a program. I like to be kept informed about everything that's happening here at Parks. How can the project office expect to get support out of the functional departments if we're kept in the dark until the very last minute? My people have been working with the existing materials for the last two months and you're telling us that it was all for nothing. Now you're giving us a material that's so new that we have no information on it whatsoever. We're now going to have to play catch-up, and that's going to cost you plenty."

One week before the 180-day milestone meeting, Gary submitted the handout package to Lord Industries for preliminary review. An hour later the phone rang.

Customer: We've just read your handout. Where did this new material come from? How come we were not informed that this work was going on? You know, of course, that our customer, the Army, will be at this meeting. How can we explain this to them? We're postponing the review meeting until all of our people have analyzed the data and are prepared to make a decision.

The purpose of a review or interchange meeting is to exchange information when *both* parties have familiarity with the topic. Normally, we (Lord Industries) require almost weekly interchange meetings with our other customers because we don't trust them. We've disregarded this policy with Parks Corporation based on past working relationships. But with the new state of developments, you have forced us to revert to our previous position, since we now question Parks Corporation's integrity in communicating with us. At first we believed this was due to an inexperienced program manager. Now, we're not sure.

Anderson: I wonder if the real reason we have these interchange meetings isn't to show our people that Lord Industries doesn't trust us. You're creating a hell of a lot of work for us, you know.

Customer: You people put yourself in this position. Now you have to live with it.

Two weeks later Lord reluctantly agreed that the new material offered the greatest promise. Three weeks later the design review meeting was held. The Army was definitely not pleased with the prime contractor's recommendation to put a new, untested material into a multimillion-dollar effort.

The Communications Breakdown

During the week following the design review meeting Gary planned to make the first verification mix in order to establish final specifications for selection of the raw materials. Unfortunately, the manufacturing plans were a week behind schedule, primarily because of Gary, since he had decided to reduce costs by accepting the responsibility for developing the bill of materials himself.

A meeting was called by Gary to consider rescheduling of the mix.

Anderson: As you know we're about a week to ten days behind schedule. We'll have to reschedule the verification mix for late next week.

Production Manager: Our resources are committed until a month from now. You can't expect to simply call a meeting and have everything reshuffled for the Blue Spider Program. We should have been notified earlier. Engineering has the responsibility for preparing the bill of materials. Why aren't they ready?

Engineering Integration: We were never asked to prepare the bill of materials. But I'm sure that we could get it out if we work our people overtime for the next two days.

Anderson: When can we remake the mix?

Production Manager: We have to redo at least 500 sheets of paper every time we reschedule mixes. Not only that, we have to reschedule people on all three shifts. If we are to reschedule your mix, it will have to be performed on overtime. That's going to increase your costs. If that's agreeable with you, we'll try it. But this will be the first and last time that production will bail you out. There are procedures that have to be followed.

Testing Engineer: I've been coming to these meetings since we kicked off this program. I think I speak for the entire engineering division when I say that the role that the director of engineering is playing in this program is suppressing individuality among our highly competent personnel. In new projects, especially those involving R&D, our people are not apt to stick their necks out. Now our people are becoming ostriches. If they're impeded from contributing, even in their own slight way, then you'll probably lose them before the project gets completed. Right now I feel that I'm wasting my time here. All I need are minutes of the team meetings and I'll be happy. Then I won't have to come to these pretend meetings anymore.

The purpose of the verification mix was to make a full-scale production run of the material to verify that there would be no material property changes in scale-up from the small mixes made in the R&D laboratories. After testing, it became obvious that the wrong lots of raw materials were used in the production verification mix.

A meeting was called by Lord Industries for an explanation of why the mistake had occurred and what the alternatives were.

Lord: Why did the problem occur?

Anderson: Well, we had a problem with the bill of materials. The result was that the mix had to be made on overtime. And when you work people on overtime, you have to be willing to accept mistakes as being a way of life. The energy cycles of our people are slow during the overtime hours.

Lord: The ultimate responsibility has to be with you, the program manager. We, at

Lord, think that you're spending too much time doing and not enough time managing. As the prime contractor, we have a hell of a lot more at stake than you do. From now on we want documented weekly technical interchange meetings and closer interaction by our quality control section with yours.

Anderson: These additional team meetings are going to tie up our key people. I can't spare people to prepare handouts for weekly meetings with your people.

Lord: Team meetings are a management responsibility. If Parks does not want the Blue Spider Program, I'm sure we can find another subcontractor. All you (Gary) have to do is give up taking the material vendors to lunch and you'll have plenty of time for handout preparation.

Gary left the meeting feeling as though he had just gotten raked over the coals. For the next two months, Gary worked sixteen hours a day, almost every day. Gary did not want to burden his staff with the responsibility of the handouts, so he began preparing them himself. He could have hired additional staff, but with such a tight budget, and having to remake verification mix, cost overruns appeared inevitable.

As the end of the seventh month approached, Gary was feeling pressure from within Parks Corporation. The decision-making process appeared to be slowing down and Gary found it more and more difficult to motivate his people. In fact, the grapevine was referring to the Blue Spider Project as a loser, and some of his key people acted as though they were on a sinking ship.

By the time the eighth month rolled around, the budget had nearly been expended. Gary was tired of doing everything himself. "Perhaps I should have stayed an engineer," thought Gary. Elliot Grey and Gary Anderson had a meeting to see what could be salvaged. Grey agreed to get Gary additional corporate funding to complete the project. "But performance must be met, since there is a lot riding on the Blue Spider Project," asserted Grey. He called a team meeting to identify the program status.

Anderson: It's time to map out our strategy for the remainder of the program. Can engineering and production adhere to the schedule that I have laid out before you?

Team Member: Engineering: This is the first time that I've seen this schedule. You can't expect me to make a decision in the next ten minutes and commit the resources of my department. We're getting a little unhappy being kept in the dark until the last minute. What happened to effective planning?

Anderson: We still have effective planning. We must adhere to the original schedule, or at least try to adhere to it. This revised schedule will do that.

Team Member: Engineering: Look, Gary! When a project gets in trouble it is usually the functional departments that come to the rescue. But if we're kept in the dark, then how can you expect us to come to your rescue? My boss wants to know, well in advance, every decision that you're contemplating with regard to our departmental resources. Right now, we . . . .

Anderson: Granted, we may have had a communications problem. But now we're in trouble and have to unite forces. What is your impression as to whether your department can meet the new schedule?

Team Member: Engineering: When the Blue Spider Program first got in trouble, my boss exercised his authority to make all departmental decisions regarding the program himself. I'm just a puppet. I have to check with him on everything.

Team Member: Production: I'm in the same boat, Gary. You know we're not happy having to reschedule our facilities and people. We went through this once before. I also have to check with my boss before giving you an answer about the new schedule.

The following week the verification mix was made. Testing proceeded according to the revised schedule, and it looked as though the total schedule milestones could be met, provided that specifications could be adhered to.

Because of the revised schedule, some of the testing had to be performed on holidays. Gary wasn't pleased with asking people to work on Sundays and holidays, but he had no choice, since the test matrix called for testing to be accomplished at specific times after end-of-mix.

A team meeting was called on Wednesday to resolve the problem of who would work on the holiday, which would occur on Friday, as well as staffing Saturday and Sunday. During the team meeting Gary became quite disappointed. Phil Rodgers, who had been Gary's test engineer since the project started, was assigned to a new project that the grapevine called Gable's new adventure. His replacement was a relatively new man, only eight months with the company. For an hour and a half, the team members argued about the little problems and continually avoided the major question, stating that they would first have to coordinate commitments with their bosses. It was obvious to Gary that his team members were afraid to make major decisions and therefore "ate up" a lot of time on trivial problems.

On the following day, Thursday, Gary went to see the department manager responsible for testing, in hopes that he could use Phil Rodgers this weekend.

Department Manager: I have specific instructions from the boss (director of engineering) to use Phil Rodgers on the new project. You'll have to see the boss if you want him back

Anderson: But we have testing that must be accomplished this weekend. Where's the new man you assigned yesterday?

Department Manager: Nobody told me you had testing scheduled for this weekend. Half of my department is already on an extended weekend vacation, including Phil Rodgers and the new man. How come I'm always the last to know when we have a problem?

Anderson: The customer is flying down his best people to observe this weekend's tests. It's too late to change anything. You and I can do the testing.

Department Manager: Not on your life. I'm staying as far away as possible from the Blue Spider Project. I'll get you someone, but it won't be me. That's for sure!

The weekend's testing went according to schedule. The raw data was made available to the customer under the stipulation that the final company position would be announced at the end of the next month, after the functional departments had a chance to analyze it.

Final testing was completed during the second week of the ninth month. The initial results looked excellent. The materials were within contract specifications, and although they were new, both Gary and Lord's management felt that there would be little difficulty in convincing the Army that this was the way to go. Henry Gable visited Gary and congratulated him on a job well done.

All that now remained was the making of four additional full-scale verification mixes in order to determine how much deviation there would be in material properties between full-sized production-run mixes. Gary tried to get the customer to concur (as part of the original trade-off analysis) that two of the four production runs could be deleted. Lord's management refused, insisting that contractual requirements must be met at the expense of the contractor.

The following week, Elliot Grey called Gary in for an emergency meeting concerning expenditures to date.

Elliot Grey: Gary, I just received a copy of the financial planning report for last quarter in which you stated that both the cost and performance of the Blue Spider Project were 75 percent complete. I don't think you realize what you've done. The target profit on the program was $200,000. Your memo authorized the vice president and general manager to book 75 percent of that, or $150,000, for corporate profit spending for stockholders. I was planning on using all $200,000 together with the additional $300,000 I personally requested from corporate headquarters to bail you out. Now I have to go back to the vice president and general manager and tell them that we've made a mistake and that we'll need an additional $150,000.

Anderson: Perhaps I should go with you and explain my error. Obviously, I take all responsibility.

Grey: No, Gary. It's our error, not yours. I really don't think you want to be around the general manager when he sees red at the bottom of the page. It takes an act of God to get money back once corporate books it as profit. Perhaps you should reconsider project engineering as a career instead of program management. Your performance hasn't exactly been sparkling, you know.

Gary returned to his office quite disappointed. No matter how hard he worked, the bureaucratic red tape of project management seemed always to do him in. But late that afternoon, Gary's disposition improved. Lord Industries called to say that, after consultation with the Army, Parks Corporation would be awarded a sole-source contract for qualification and production of Spartan missile components using the new longer-life raw materials. Both Lord and the Army felt that the sole-source contract was justified, provided that continued testing showed the same results, since Parks Corporation had all of the technical experience with the new materials.

Gary received a letter of congratulations from corporate headquarters, but no additional pay increase. The grapevine said that a substantial bonus was given to the director of engineering.

During the tenth month, results were coming back from the accelerated aging tests performed on the new materials. The results indicated that, although the new materials would meet specifications, the age life would probably be less than five years. These numbers came as a shock to Gary. Gary and Paul Evans had a conference to determine the best strategy to follow.

Anderson: Well, I guess we're now in the fire instead of the frying pan. Obviously, we can't tell Lord Industries about these tests. We ran them on our own. Could the results be wrong?

Evans: Sure, but I doubt it. There's always margin for error when you perform accelerated aging tests on new materials. There can be reactions taking place that we know nothing about. Furthermore, the accelerated aging tests may not even correlate well with actual aging. We must form a company position on this as soon as possible.

Anderson: I'm not going to tell anyone about this, especially Henry Gable. You and I will handle this. It will be my throat if word of this leaks out. Let's wait until we have the production contract in hand.

Evans: That's dangerous. This has to be a company position, not a project office position. We had better let them know upstairs.

Anderson: I can't do that. I'll take all responsibility. Are you with me on this?

Evans: I'll go along. I'm sure I can find employment elsewhere when we open Pandora's box. You had better tell the department managers to be quiet also.

Two weeks later, as the program was winding down into the testing for the final verification mix and final report development, Gary received an urgent phone call asking him to report immediately to Henry Gable's office.

Gable: When this project is over, you're through. You'll never hack it as a program manager, or possibly a good project engineer. We can't run projects around here without honesty and open communications. How the hell do you expect top management to support you when you start censoring bad news to the top? I don't like surprises. I like to get the bad news from the program manager and project engineers, not secondhand from the customer. And of course, we cannot forget the cost overrun. Why didn't you take some precautionary measures?

Anderson: How could I when you were asking our people to do work such as accelerated aging tests that would be charged to my project and was not part of program plan? I don't think I'm totally to blame for what's happened.

Gable: Gary, I don't think it's necessary to argue the point any further. I'm willing to give you back your old job, in engineering. I hope you didn't lose too many friends while working in program management. Finish up final testing and the program report. Then I'll reassign you.

Gary returned to his office and put his feet up on the desk. "Well," thought Gary, "perhaps I'm better off in engineering. At least I can see my wife and kids once in a while." As Gary began writing the final report, the phone rang:

Functional Manager: Hello, Gary. I just thought I'd call to find out what charge number you want us to use for experimenting with this new procedure to determine accelerated age life.

Anderson: Don't call me! Call Gable. After all, the Blue Spider Project is his baby.

Questions

1. If you were Gary Anderson, would you have accepted this position after the director stated that this project would be his baby all the way?

2. Do engineers with MBA degrees aspire to high positions in management?
3. Was Gary qualified to be a project manager?
4. What are the moral and ethical issues facing Gary?
5. What authority does Gary Anderson have and to whom does he report?
6. Is it true when you enter project management, you either go up the organization or out the door?
7. Is it possible for an executive to take too much of an interest in an R&D project?
8. Should Paul Evans have been permitted to report information to Gable before reporting it to the project manager?
9. Is it customary for the project manager to prepare all of the handouts for a customer interchange meeting?
10. What happens when a situation of mistrust occurs between the customer and contractor?
11. Should functional employees of the customer and contractor be permitted to communicate with one another without going through the project office?
12. Did Gary demonstrate effective time management?
13. Did Gary understand production operations?
14. Are functional employees authorized to make project decisions?
15. On R&D projects, should profits be booked periodically or at project termination?
16. Should a project manager ever censor bad news?
17. Could the above-mentioned problems have been resolved if there had been a singular methodology for project management in place?
18. Can a singular methodology for project management specify morality and ethics in dealing with customers? If so, how do we then handle situations where the project manager violates protocol?
19. Could the lessons learned on success and failure during project debriefings cause a major change in the project management methodology?

Case 15

Corwin Corporation*

By June 1983, Corwin Corporation had grown into a $150 million per year corporation with an international reputation for manufacturing low-cost, high-quality rubber components. Corwin maintained more than a dozen different product lines, all of which were sold as off-the-shelf items in department stores, hardware stores, and automotive parts distributors. The name "Corwin" was now synonymous with "quality." This provided management with the luxury of having products that maintained extremely long life cycles.

Organizationally, Corwin had maintained the same structure for more than 15 years (see Exhibit I). The top management of Corwin Corporation was highly conservative and believed in using a marketing approach to find new markets for existing product lines rather than exploring for new products. Under this philosophy, Corwin maintained a small R&D group whose mission was simply to evaluate state-of-the-art technology and its application to existing product lines.

Corwin's reputation was so good that they continually received inquiries about the manufacturing of specialty products. Unfortunately, the conservative nature of Corwin's management created a "do not rock the boat" atmosphere opposed to taking any type of risks. A management policy was established to evaluate all specialty-product requests. The policy required answering yes to the following questions:

- Will the specialty product provide the same profit margin (20 percent) as existing product lines?
- What is the total projected profitability to the company in terms of follow-on contracts?

*Reprinted from H. Kerzner, *Project Management: A Systems Approach to Planning, Scheduling and Controlling,* 6th ed. New York: Wiley, 1998, pp. 509–517.

Exhibit I. Organizational chart for Corwin Corporation

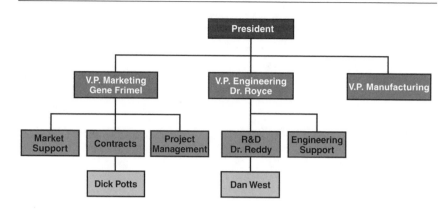

- Can the specialty product be developed into a product line?
- Can the specialty product be produced with minimum disruption to existing product lines and manufacturing operations?

These stringent requirements forced Corwin to no-bid more than 90 percent of all specialty-product inquiries.

Corwin Corporation was a marketing-driven organization, although manufacturing often had different ideas. Almost all decisions were made by marketing with the exception of product pricing and estimating, which was a joint undertaking between manufacturing and marketing. Engineering was considered as merely a support group to marketing and manufacturing.

For specialty products, the project managers would always come out of marketing, even during the R&D phase of development. The company's approach was that if the specialty product should mature into a full product line, then there should be a product line manager assigned right at the onset.

The Peters Company Project

In 1980, Corwin accepted a specialty-product assignment from Peters Company because of the potential for follow-on work. In 1981 and 1982, and again in 1983, profitable follow-on contracts were received, and a good working relationship developed, despite Peters' reputation for being a difficult customer to work with.

On December 7, 1982, Gene Frimel, the vice president of marketing at Corwin, received a rather unusual phone call from Dr. Frank Delia, the marketing vice president at Peters Company.

Frank Delia: Gene, I have a rather strange problem on my hands. Our R&D group has $250,000 committed for research toward development of a new rubber product material, and we simply do not have the available personnel or talent to undertake the project. We have to go outside. We'd like your company to do the work. Our testing and R&D facilities are already overburdened.

Gene Frimel: Well, as you know, Frank, we are not a research group even though we've done this once before for you. And furthermore, I would never be able to sell our management on such an undertaking. Let some other company do the R&D work and then we'll take over on the production end.

Delia: Let me explain our position on this. We've been burned several times in the past. Projects like this generate several patents, and the R&D company almost always requires that our contracts give them royalties or first refusal for manufacturing rights.

Frimel: I understand your problem, but it's not within our capabilities. This project, if undertaken, could disrupt parts of our organization. We're already operating lean in engineering.

Delia: Look, Gene! The bottom line is this: We have complete confidence in your manufacturing ability to such a point that we're willing to commit to a five-year production contract if the product can be developed. That makes it extremely profitable for you.

Frimel: You've just gotten me interested. What additional details can you give me?

Delia: All I can give you is a rough set of performance specifications that we'd like to meet. Obviously, some trade-offs are possible.

Frimel: When can you get the specification sheet to me?

Delia: You'll have it tomorrow morning. I'll ship it overnight express.

Frimel: Good! I'll have my people look at it, but we won't be able to get you an answer until after the first of the year. As you know, our plant is closed down for the last two weeks in December, and most of our people have already left for extended vacations.

Delia: That's not acceptable! My management wants a signed, sealed, and delivered contract by the end of this month. If this is not done, corporate will reduce our budget for 1983 by $250,000, thinking that we've bitten off more than we can chew. Actually, I need your answer within 48 hours so that I'll have some time to find another source if necessary.

Frimel: You know, Frank, today is December 7, Pearl Harbor Day. Why do I feel as though the sky is about to fall in?

Delia: Don't worry, Gene! I'm not going to drop any bombs on you. Just remember, all that we have available is $250,000, and the contract must be a firm-fixed-price effort. We anticipate a six-month project with $125,000 paid on contract signing and the balance at project termination.

Frimel: I still have that ominous feeling, but I'll talk to my people. You'll hear from us with a go or no-go decision within 48 hours. I'm scheduled to go on a cruise in the Caribbean, and my wife and I are leaving this evening. One of my people will get back to you on this matter.

Gene Frimel had a problem. All bid and no-bid decisions were made by a four-man committee composed of the president and the three vice presidents. The president and the vice president for manufacturing were on vacation. Frimel met with Dr. Royce, the vice president of engineering, and explained the situation.

Royce: You know, Gene, I totally support projects like this because it would help our technical people grow intellectually. Unfortunately, my vote never appears to carry any weight.

Frimel: The profitability potential, as well as the development of good customer relations, makes this attractive, but I'm not sure we want to accept such a risk. A failure could easily destroy our good working relationship with Peters Company.

Royce: I'd have to look at the specification sheets before assessing the risks, but I would like to give it a shot.

Frimel: I'll try to reach our president by phone.

By late afternoon, Frimel was fortunate enough to be able to contact the president and received a reluctant authorization to proceed. The problem now was how to prepare a proposal within the next two or three days and be prepared to make an oral presentation to Peters Company.

Frimel: The Boss gave his blessing, Royce, and the ball is in your hands. I'm leaving for vacation, and you'll have total responsibility for the proposal and presentation. Delia wants the presentation this weekend. You should have his specification sheets tomorrow morning.

Royce: Our R&D director, Dr. Reddy, left for vacation this morning. I wish he were here to help me price out the work and select the project manager. I assume that, in this case, the project manager will come out of engineering rather than marketing.

Frimel: Yes, I agree. Marketing should not have any role in this effort. It's your baby all the way. And as for the pricing effort, you know our bid will be for $250,000. Just work backwards to justify the numbers. I'll assign one of our contracting people to assist you in the pricing. I hope I can find someone who has experience in this type of effort. I'll call Delia and tell him we'll bid it with an unsolicited proposal.

Royce selected Dan West, one of the R&D scientists, to act as the project leader. Royce had severe reservations about doing this without the R&D director, Dr. Reddy, being actively involved. With Reddy on vacation, Royce had to make an immediate decision.

On the following morning, the specification sheets arrived and Royce, West, and Dick Potts, a contracts man, began preparing the proposal. West prepared the direct labor man-hours, and Royce provided the costing data and pricing rates. Potts, being completely unfamiliar with this type of effort, simply acted as an observer and provided legal advice when necessary. Potts allowed Royce to make all decisions even though the contracts man was considered the official representative of the president.

Finally completed two days later, the proposal was actually a ten-page letter that simply contained the cost summaries (see Exhibit II) and the engineering intent. West estimated that 30 tests would be required. The test matrix described the test conditions only for the first five tests. The remaining 25 test conditions would be determined at a later date, jointly by Peters and Corwin personnel.

Exhibit II. Proposal cost summaries

Direct labor and support	$ 30,000
Testing (30 tests at $2,000 each)	60,000
Overhead at 100%	90,000
Materials	30,000
G&A (general and administrative, 10%)	21,000
Total	$231,000
Profit	19,000
Total	$250,000

On Sunday morning, a meeting was held at Peters Company, and the proposal was accepted. Delia gave Royce a letter of intent authorizing Corwin Corporation to begin working on the project immediately. The final contract would not be available for signing until late January, and the letter of intent simply stated that Peters Company would assume all costs until such time that the contract was signed or the effort terminated.

West was truly excited about being selected as the project manager and being able to interface with the customer, a luxury that was usually given only to the marketing personnel. Although Corwin Corporation was closed for two weeks over Christmas, West still went into the office to prepare the project schedules and to identify the support he would need in the other areas, thinking that if he presented this information to management on the first day back to work, they would be convinced that he had everything under control.

The Work Begins

On the first working day in January 1983, a meeting was held with the three vice presidents and Dr. Reddy to discuss the support needed for the project. (West was not in attendance at this meeting, although all participants had a copy of his memo.)

Reddy: I think we're heading for trouble in accepting this project. I've worked with Peters Company previously on R&D efforts, and they're tough to get along with. West is a good man, but I would never have assigned him as the project leader. His expertise is in managing internal rather than external projects. But, no matter what happens, I'll support West the best I can.

Royce: You're too pessimistic. You have good people in your group and I'm sure you'll be able to give him the support he needs. I'll try to look in on the project every so often. West will still be reporting to you for this project. Try not to burden him too much with other work. This project is important to the company.

West spent the first few days after vacation soliciting the support that he needed from the other line groups. Many of the other groups were upset that they had not been informed earlier and were unsure as to what support they could provide. West met with Reddy to discuss the final schedules.

Reddy: Your schedules look pretty good, Dan. I think you have a good grasp on the problem. You won't need very much help from me. I have a lot of work to do on other activities, so I'm just going to be in the background on this project. Just drop me a note every once in a while telling me what's going on. I don't need anything formal. Just a paragraph or two will suffice.

By the end of the third week, all of the raw materials had been purchased, and initial formulations and testing were ready to begin. In addition, the contract was ready for signature. The contract contained a clause specifying that Peters Company had the right to send an in-house representative into Corwin Corporation for the duration of the project. Peters Company informed Corwin that Patrick Ray would be the in-house representative, reporting to Delia, and would assume his responsibilities on or about February 15.

By the time Pat Ray appeared at Corwin Corporation, West had completed the first three tests. The results were not what was expected, but gave promise that Corwin was heading in the right direction. Pat Ray's interpretation of the tests was completely opposite to that of West. Ray thought that Corwin was "way off base," and that redirection was needed.

Pat Ray: Look, Dan! We have only six months to do this effort and we shouldn't waste our time on marginally acceptable data. These are the next five tests I'd like to see performed.

Dan West: Let me look over your request and review it with my people. That will take a couple of days, and, in the meanwhile, I'm going to run the other two tests as planned.

Ray's arrogant attitude bothered West. However, West decided that the project was too important to "knock heads" with Ray and simply decided to cater to Ray the best he could. This was not exactly the working relationship that West expected to have with the in-house representative.

West reviewed the test data and the new test matrix with engineering personnel, who felt that the test data was inconclusive as yet and preferred to withhold their opinion until the results of the fourth and fifth tests were made available. Although this displeased Ray, he agreed to wait a few more days if it meant getting Corwin Corporation on the right track.

The fourth and fifth tests appeared to be marginally acceptable just as the first three had been. Corwin's engineering people analyzed the data and made their recommendations.

West: Pat, my people feel that we're going in the right direction and that our path has greater promise than your test matrix.

Ray: As long as we're paying the bills, we're going to have a say in what tests are conducted. Your proposal stated that we would work together in developing the other test conditions. Let's go with my test matrix. I've already reported back to my boss that the first five tests were failures and that we're changing the direction of the project.

West: I've already purchased $30,000 worth of raw materials. Your matrix uses other materials and will require additional expenditures of $12,000.

Ray: That's your problem. Perhaps you shouldn't have purchased all of the raw materials until we agreed on the complete test matrix.

During the month of February, West conducted 15 tests, all under Ray's direction. The tests were scattered over such a wide range that no valid conclusions could be drawn. Ray continued sending reports back to Delia confirming that Corwin was not producing beneficial results and there was no indication that the situation would reverse itself. Delia ordered Ray to take any steps necessary to ensure a successful completion of the project.

Ray and West met again as they had done for each of the past 45 days to discuss the status and direction of the project.

Ray: Dan, my boss is putting tremendous pressure on me for results, and thus far I've given him nothing. I'm up for promotion in a couple of months and I can't let this project stand in my way. It's time to completely redirect the project.

West: Your redirection of the activities is playing havoc with my scheduling. I have people in other departments who just cannot commit to this continual rescheduling. They blame me for not communicating with them when, in fact, I'm embarrassed to.

Ray: Everybody has their problems. We'll get this problem solved. I spent this morning working with some of your lab people in designing the next 15 tests. Here are the test conditions.

West: I certainly would have liked to be involved with this. After all, I thought I was the project manager. Shouldn't I have been at the meeting?

Ray: Look, Dan! I really like you, but I'm not sure that you can handle this project. We need some good results immediately, or my neck will be stuck out for the next four months. I don't want that. Just have your lab personnel start on these tests, and we'll get along fine. Also, I'm planning on spending a great deal of time in your lab area. I want to observe the testing personally and talk to your lab personnel.

West: We've already conducted 20 tests, and you're scheduling another 15 tests. I priced out only 30 tests in the proposal. We're heading for a cost overrun condition.

Ray: Our contract is a firm-fixed-price effort. Therefore, the cost overrun is your problem.

West met with Dr. Reddy to discuss the new direction of the project and potential cost overruns. West brought along a memo projecting the costs through the end of the third month of the project (see Exhibit III).

Reddy: I'm already overburdened on other projects and won't be able to help you out. Royce picked you to be the project manager because he felt that you could do the job. Now, don't let him down. Send me a brief memo next month explaining the situation, and I'll see what I can do. Perhaps the situation will correct itself.

During the month of March, the third month of the project, West received almost daily phone calls from the people in the lab stating that Pat Ray was interfering with their job. In fact, one phone call stated that Ray had changed the test conditions from what was agreed on in the latest test matrix. When West confronted Ray on his med-

Exhibit III. Projected cost summary at the end of the third month

	Original Proposal Cost Summary for Six-Month Project	Total Project Costs Projected at End of Third Month
Direct labor/support	$ 30,000	$ 15,000
Testing	60,000 (30 tests)	70,000 (35 tests)
Overhead	90,000 (100%)	92,000 (120%)*
Materials	30,000	50,000
G&A	21,000 (10%)	22,700 (10%)
Totals	$231,000	$249,700

*Total engineering overhead was estimated at 100 percent, whereas the R&D overhead was 120 percent.

dling, Ray asserted that Corwin personnel were very unprofessional in their attitude and that he thought this was being carried down to the testing as well. Furthermore, Ray demanded that one of the functional employees be removed immediately from the project because of incompetence. West stated that he would talk to the employee's department manager. Ray, however, felt that this would be useless and said, "Remove him or else!" The functional employee was removed from the project.

By the end of the third month, most Corwin employees were becoming disenchanted with the project and were looking for other assignments. West attributed this to Ray's harassment of the employees. To aggravate the situation even further, Ray met with Royce and Reddy, and demanded that West be removed and a new project manager be assigned.

Royce refused to remove West as project manager, and ordered Reddy to take charge and help West get the project back on track.

Reddy: You've kept me in the dark concerning this project, West. If you want me to help you, as Royce requested, I'll need all the information tomorrow, especially the cost data. I'll expect you in my office tomorrow morning at 8:00 A.M. I'll bail you out of this mess.

West prepared the projected cost data for the remainder of the work and presented the results to Dr. Reddy (see Exhibit IV). Both West and Reddy agreed that the project was now out of control, and severe measures would be required to correct the situation, in addition to more than $250,000 in corporate funding.

Reddy: Dan, I've called a meeting for 10:00 A.M. with several of our R&D people to completely construct a new test matrix. This is what we should have done right from the start.

West: Shouldn't we invite Ray to attend this meeting? I'm sure he'd want to be involved in designing the new test matrix.

Reddy: I'm running this show now, not Ray!! Tell Ray that I'm instituting new policies and procedures for in-house representatives. He's no longer authorized to visit the labs at his own discretion. He must be accompanied by either you or me. If he

Exhibit IV. Estimate of total project completion costs

Direct labor/support	$ 47,000*
Testing (60 tests)	120,000
Overhead (120%)	200,000
Materials	103,000
G&A	47,000
	$517,000
Peters contract	250,000
Overrun	$267,000

*Includes Dr. Reddy.

doesn't like these rules, he can get out. I'm not going to allow that guy to disrupt our organization. We're spending our money now, not his.

West met with Ray and informed him of the new test matrix as well as the new policies and procedures for in-house representatives. Ray was furious over the new turn of events and stated that he was returning to Peters Company for a meeting with Delia.

On the following Monday, Frimel received a letter from Delia stating that Peters Company was officially canceling the contract. The reasons given by Delia were as follows:

1. Corwin had produced absolutely no data that looked promising.
2. Corwin continually changed the direction of the project and did not appear to have a systematic plan of attack.
3. Corwin did not provide a project manager capable of handling such a project.
4. Corwin did not provide sufficient support for the in-house representative.
5. Corwin's top management did not appear to be sincerely interested in the project and did not provide sufficient executive-level support.

Royce and Frimel met to decide on a course of action in order to sustain good working relations with Peters Company. Frimel wrote a strong letter refuting all of the accusations in the Peters letter, but to no avail. Even the fact that Corwin was willing to spend $250,000 of their own funds had no bearing on Delia's decision. The damage was done. Frimel was now thoroughly convinced that a contract should not be accepted on "Pearl Harbor Day."

Questions
1. What were the major mistakes made by Corwin?
2. Should Corwin have accepted the assignment?
3. Should companies risk bidding on projects based upon rough draft specifications?
4. Should the shortness of the proposal preparation time have required more active top management involvement before the proposal went out-of-house?
5. Are there any risks in not having the vice president for manufacturing available during the go or no-go bidding decision?

6. Explain the attitude of Dick Potts during the proposal activities.

7. None of the executives expressed concern when Dr. Reddy said, "I would never have assigned him (West) as project leader." How do you account for the executives' lack of concern?

8. How important is it to inform line managers of proposal activities even if the line managers are not required to provide proposal support?

9. Explain Dr. Reddy's attitude after go-ahead.

10. How should West have handled the situation where Pat Ray's opinion of the test data was contrary to that of Corwin's engineering personnel?

11. How should West have reacted to the remarks made by Ray that he informed Delia that the first five tests were failures?

12. Is immediate procurement of all materials a mistake?

13. Should Pat Ray have been given the freedom to visit laboratory personnel at any time?

14. Should an in-house representative have the right to remove a functional employee from the project?

15. Financially, how should the extra tests have been handled?

16. Explain Dr. Reddy's attitude when told to assume control of the project.

17. Delia's letter, stating the five reasons for canceling the project, was refuted by Frimel, but with no success. Could Frimel's early involvement as a project sponsor have prevented this?

18. In retrospect, would it have been better to assign a marketing person as project manager?

19. Your company has a singular methodology for project management. You are offered a special project from a powerful customer that does not fit into your methodology. Should a project be refused simply because it is not a good fit with your methodology?

20. Should a customer be informed that only projects that fit your methodology would be accepted?

Case 16

MIS Project Management at First National Bank*

First National Bank (FNB) had been one of the fastest-growing banks in the Midwest (1970s and early 1980s). The holding company of the bank has been actively involved in purchasing small banks thoughout the state of Ohio. This expansion and the resulting increase of operations has been attended by considerable growth in numbers of employees and in the complexity of the organizational structure. In five years the staff of the bank has increased by 35 percent, and total assets have grown by 70 percent. FNB management is eagerly looking forward to a change in the Ohio banking laws that will allow statewide branch banking.

Information Services Division (ISD) History

Data processing at FNB has grown at a much faster pace than the rest of the bank. The systems and programming staff grew from 12 in 1970 to over 75 during the first part of 1977. Because of several future projects, the staff is expected to increase by 50 percent during the next two years.

Prior to 1972, the Information Services Department reported to the executive vice president of the Consumer Banking and Operations Division. As a result, the first banking applications to be computerized were in the demand deposit, savings, and consumer credit banking areas. The computer was seen as a tool to speed up the processing of consumer transactions. Little effort was expended to meet the informational requirements of the rest of the bank. This caused a high-level conflict, since each major operating organization of the bank did not have equal access to systems

---

*Reprinted from H. Kerzner, *Project Management: A Systems Approach to Planning, Scheduling and Controlling,* 6th ed. New York: Wiley, 1998, pp. 359–367.

Exhibit I.   Information Services Division organizational chart

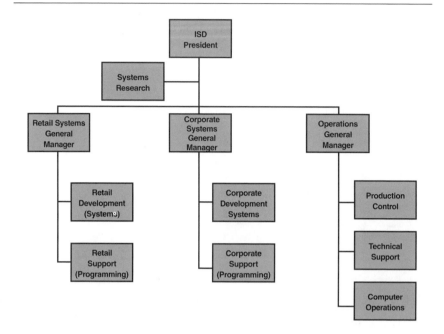

and programming resources. The management of FNB became increasingly aware of the benefits that could accrue from a realignment of the bank's organization into one that would be better attuned to the total information requirements of the corporation.

In 1982 the Information Services Division (ISD) was created. ISD was removed from the Consumer Banking and Operations Division to become a separate division reporting directly to the president. An organizational chart depicting the Information Services Division is shown in Exhibit I.

Priorities Committee

During 1982 the Priorities Committee was formed. It consists of the chief executive officer of each of the major operating organizations whose activities are directly affected by the need for new or revised information systems. The Priorities Committee was established to ensure that the resources of systems and programming personnel and computer hardware would be used only on those information systems that can best be cost justified. Divisions represented on the committee are included in Exhibit II.

The Priorities Committee meets monthly to reaffirm previously set priorities and rank new projects introduced since the last meeting. Bank policy states that the only way to obtain funds for an information development project is to submit a request to the Priorities Committee and have it approved and ranked in overall priority order for the bank. Placing potential projects in ranked sequence is done by the senior executives. The primary document used for Priorities Committee review is called the project proposal.

Exhibit II. **First National Bank organizational chart**

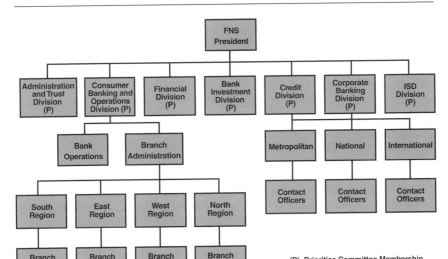

(P)=Priorities Committee Membership

The Project Proposal Life Cycle

When a user department determines a need for the development or enhancement of an information system, it is required to prepare a draft containing a statement of the problem from its functional perspective. The problem statement is sent to the president of ISD, who authorizes Systems Research (see Exhibit I) to prepare an impact statement. This impact statement will include a general overview from ISD's perspective of:

- Project feasibility
- Project complexity
- Conformity with long-range ISD plans
- Estimated ISD resource commitment
- Review of similar requests
- Unique characteristics/problems
- Broad estimate of total costs

The problem and impact statements are then presented to the members of the Priorities Committee for their review. The proposals are preliminary in nature, but they permit the broad concept (with a very approximate cost attached to it) to be reviewed by the executive group to see if there is serious interest in pursuing the idea. If the interest level of the committee is low, then the idea is rejected. However, if the Priorities Committee members feel the concept has merit, they authorize the Systems Research Group of ISD to prepare a full-scale project proposal that contains:

- A detailed statement of the problem
- Identification of alternative solutions
- Impact of request on:

- User division
- ISD
- Other operating divisions
- Estimated costs of solutions
- Schedule of approximate task duration
- Cost-benefit analysis of solutions
- Long-range implications
- Recommended course of action

After the project proposal is prepared by systems research, the user sponsor must review the proposal and appear at the next Priorities Committee meeting to speak in favor of the approval and priority level of the proposed work. The project proposal is evaluated by the committee and either dropped, tabled for further review, or assigned a priority relative to ongoing projects and available resources.

The final output of a Priorities Committee meeting is an updated list of project proposals in priority order with an accompanying milestone schedule that indicates the approximate time span required to implement each of the proposed projects.

The net result of this process is that the priority-setting for systems development is done by a cross section of executive management; it does not revert by default to data processing management. Priority-setting, if done by data processing, can lead to misunderstanding and dissatisfaction by sponsors of the projects that did not get ranked high enough to be funded in the near future. The project proposal cycle at FNB is diagrammed in Exhibit III. Once a project has risen to the top of the ranked priority list, it is assigned to the appropriate systems group for systems definition, system design and development, and system implementation.

The time spent by systems research in producing impact statements and project proposals is considered to be overhead by ISD. No systems research time is directly charged to the development of information systems.

Project Life Cycle

As noted before, the systems and programming staff of ISD has increased in size rapidly and is expected to expand by another 50 percent over the next two years. As a rule, most new employees have previous data processing experience and training in various systems methodologies. ISD management recently implemented a project management system dedicated to providing a uniform step-by-step methodology for the development of management information systems. All project work is covered by tasks that make up the information project development life cycle at FNB. The subphases used by ISD in the project life cycle are:

1. Systems definition
   a. Project plan
   b. User requirements
   c. Systems definition
   d. Advisability study
2. Systems design and development
   a. Preliminary systems design
   b. Subsystems design
   c. Program design
   d. Programming and testing

3. System implementation
    a. System implementation
    b. System test
    c. Production control turnover
    d. User training
    e. System acceptance

Project Estimating

The project management system contains a list of all normal tasks and subtasks (over 400) to be performed during the life cycle of a development project. The project manager must examine all the tasks to determine if they apply to a given project. The manager must insert additional tasks if required and delete tasks that do not apply. The project manager next estimates the amount of time (in hours) to complete each task of each subphase of the project life cycle.

The estimating process of the project management system uses a "moving window" concept. ISD management feels that detailed cost estimating and time schedules are only meaningful for the next subphase of a project, where the visibility of the tasks to be performed is quite clear. Beyond that subphase, a more summary method of estimating is relied on. As the project progresses, new segments of the project gain visibility. Detailed estimates are made for the next major portion of the project, and summary estimates are done beyond that until the end of the project.

Estimates are performed at five intervals during the project life cycle. When the project is first initiated, the funding is based on the original estimates, which are derived from the list of normal tasks and subtasks. At this time, the subphases through the advisability study are estimated in detail, and summary estimates are prepared for the rest of the tasks in the project. Once the project has progressed through the advisability study, the preliminary systems design is estimated in detail, and the balance of the project is estimated in a more summary fashion. Estimates are conducted in this manner until the systems implementation plan is completed and the scope of the remaining subphases of the project is known. This multiple estimating process is used because it is almost impossible at the beginning of many projects to be certain of what the magnitude of effort will be later on in the project life cycle.

Funding of Projects

The project plan is the official document for securing funding from the sponsor in the user organization. The project plan must be completed and approved by the project manager before activity can begin on the user requirements subphase (1b). An initial stage in developing a project plan includes the drawing of a network that identifies each of the tasks to be done in the appropriate sequence for their execution. The project plan must include a milestone schedule, a cost estimate, and a budget request. It is submitted to the appropriate general manager of systems and programming for review so that an understanding can be reached of how the estimates were prepared and why the costs and schedules are as shown. At this time the general manager can get an idea of the quantity of systems and programming resources required by the project. The general manager next sets up a meeting with the project manager and

Exhibit III. The project proposal cycle

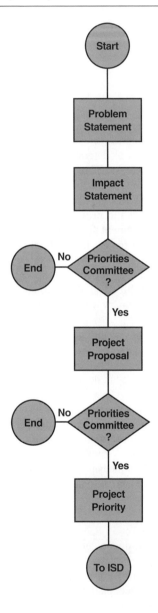

the user sponsor to review the project plan and obtain funding from the user organization.

The initial project funding is based on an estimate that includes a number of assumptions concerning the scope of the project. Once certain key milestones in the project have been achieved, the visibility on the balance of the project becomes much clearer, and reestimates are performed. The reestimates may result in refunding if there has been a significant change in the project. The normal milestone refunding points are as follows:

1. After the advisability study (1d)
2. After the preliminary systems design (2a)
3. After the program design (2c)
4. After system implementation (3a)

The refunding process is similar to the initial funding with the exception that progress information is presented on the status of the work and reasons are given to explain deviations from project expenditure projections. A revised project plan is prepared for each milestone refunding meeting.

During the systems design and development stage, design freezes are issued by the project manager to users announcing that no additional changes will be accepted to the project beyond that point. The presence of these design freezes is outlined at the beginning of the project. Following the design freeze, no additional changes will be accepted unless the project is reestimated at a new level and approved by the user sponsor.

System Quality Reviews

The key element in ensuring user involvement in the new system is the conducting of quality reviews. In the normal system cycles at FNB, there are ten quality reviews, seven of which are participated in jointly by users and data processing personnel, and three of which are technical reviews by data processing (DP) personnel only. An important side benefit of this review process is that users of a new system are forced to become involved in and are permitted to make a contribution to the systems design.

Each of the quality review points coincides with the end of a subphase in the project life cycle. The review must be held at the completion of one subphase to obtain authorization to begin work on the tasks of the next subphase of the project.

All tasks and subtasks assigned to members of the project team should end in some "deliverable" for the project documentation. The first step in conducting a quality review is to assemble the documentation produced during the subphase for distribution to the Quality Review Board. The Quality Review Board consists of between two and eight people who are appointed by the project manager with the approval of the project sponsor and the general manager of systems and programming. The minutes of the quality review meeting are written either to express "concurrence" with the subsystem quality or to recommend changes to the system that must be completed before the next subphase can be started. By this process the system is fine-tuned to the requirements of the members of the review group at the end of each subphase in the system. The members of the Quality Review Board charge their time to the project budget.

Quality review points and review board makeup are as follows:

Review	Review Board
User requirements	User oriented
Systems definition	User oriented
Advisability study	User oriented
Preliminary systems design	User oriented

Subsystems design	Users and DP
Program design	DP
Programming and testing	DP
System implementation	User oriented
System test	User oriented
Production control turnover	DP

To summarize, the quality review evaluates the quality of project subphase results, including design adequacy and proof of accomplishment in meeting project objectives. The review board authorizes work to progress based on their detailed knowledge that all required tasks and subtasks of each subphase have been successfully completed and documented.

Project Team Staffing

Once a project has risen to the top of the priority list, the appropriate manager of systems development appoints a project manager from his or her staff of analysts. The project manager has a short time to review the project proposal created by systems research before developing a project plan. The project plan must be approved by the general manager of systems and programming and the user sponsor before the project can be funded and work started on the user requirements subphase.

The project manager is "free" to spend as much time as required in reviewing the project proposal and creating the project plan; however, this time is "charged" to the project at a rate of $26 per hour. The project manager must negotiate with a "supervisor," the manager of systems development, to obtain the required systems analysts for the project, starting with the user requirements subphase. The project manager must obtain programming resources from the manager of systems support. Schedule delays caused by a lack of systems or programming resources are to be communicated to the general manager by the project manager. All ISD personnel working on a project charge their time at a rate of $26 per hour. All computer time is billed at a rate of $64 per hour.

There are no user personnel on the project team; all team members are from ISD.

Corporate Database

John Hart had for several years seen the need to use the computer to support the corporate marketing effort of the bank. Despite the fact that the majority of the bank's profits were from corporate customers, most information systems effort was directed at speeding up transactions handling for small unprofitable accounts.

Mr. Hart had extensive experience in the Corporate Banking Division of the bank. He realized the need to consolidate information about corporate customers from many areas of the bank into one corporate database. From this information corporate banking services could be developed not only to better serve the corporate customers, but also to contribute heavily to the profit structure of the bank through repricing of services.

The absence of a corporate database meant that no one individual knew what total banking services a corporate customer was using, because corporate services were provided by many banking departments. It was also impossible to determine how profitable a corporate customer was to the bank. Contact officers did not have regularly scheduled calls. They serviced corporate customers almost on a hit-or-miss basis. Unfortunately, many customers were "sold" on a service because they walked in the door and requested it. Mr. Hart felt that there was a vast market of untapped corporate customers in Ohio who would purchase services from the bank if they were contacted and "sold" in a professional manner. A corporate database could be used to develop corporate profiles to help contact officers sell likely services to corporations.

Mr. Hart knew that data about corporate customers was being processed in many departments of the bank, but mainly in the following divisions:

- Corporate Banking
- Corporate Trust
- Consumer Banking

He also realized that much of the information was processed in manual systems, some was processed by time-sharing at various vendors, and other information was computerized in many internal information systems.

The upper management of FNB must have agreed with Mr. Hart because in December of 1986 the Corporate Marketing Division was formed with John Hart as its executive vice president. Mr. Hart was due to retire within the year but was honored to be selected for the new position. He agreed to stay with the bank until "his" new system was "off the ground." He immediately composed a problem statement and sent it to the ISD. Systems Research compiled a preliminary impact statement. At the next Priorities Committee meeting, a project proposal was authorized to be done by Systems Research.

The project proposal was completed by Systems Research in record time. Most information was obtained from Mr. Hart. He had been thinking about the systems requirements for years and possessed vast experience in almost all areas of the bank. Other user divisions and departments were often "too busy" when approached for information. A common reply to a request for information was, "That project is John's baby; he knows what we need."

The project proposal as prepared by Systems Research recommended the following:

- Interfaces should be designed to extract information from existing computerized systems for the corporate database (CDB).
- Time-sharing systems should be brought in-house to be interfaced with the CDB.
- Information should be collected from manual systems to be integrated into the CDB on a temporary basis.
- Manual systems should be consolidated and computerized, potentially causing a reorganization of some departments.
- Information analysis and flow for all departments and divisions having contact with corporate customers should be coordinated by the Corporate Marketing Division.
- All corporate database analysis should be done by the Corporate Marketing Division staff, using either a user-controlled report writer or interactive inquiry.

The project proposal was presented at the next Priorities Committee meeting, where it was approved and rated as the highest priority MIS development project in the bank. Mr. Hart became the user sponsor for the CDB project.

The project proposal was sent to the manager of corporate development, who appointed Jim Gunn as project manager from the staff of analysts in corporate development. Jim Gunn was the most experienced project manager available. His prior experience consisted of successful projects in the Financial Division of the bank.

Jim reviewed the project proposal and started to work on his project plan. He was aware that the corporate analyst group was presently understaffed but was assured by his manager, the manager of corporate development, that resources would be available for the user requirements subphase. He had many questions concerning the scope of the project and the interrelationship between the Corporate Marketing Division and the other users of corporate marketing data. But each meeting with Mr. Hart ended with the same comment: "This is a waste of time. I've already been over this with Systems Research. Let's get moving." Jim also was receiving pressure from the general manager to "hurry up" with the project plan. Jim therefore quickly prepared his project plan, which included a general milestone schedule for subphase completion, a general cost estimate, and a request for funding. The project plan was reviewed by the general manager and signed by Mr. Hart.

Jim Gunn anticipated the need to have four analysts assigned to the project and went to his manager to see who was available. He was told that two junior analysts were available now and another analyst should be free next week. No senior analysts were available. Jim notified the general manager that the CDB schedule would probably be delayed because of a lack of resources, but received no response.

Jim assigned tasks to the members of the team and explained the assignments and the schedule. Since the project was understaffed, Jim assigned a heavy load of tasks to himself.

During the next two weeks the majority of the meetings set up to document user requirements were canceled by the user departments. Jim notified Mr. Hart of the problem and was assured that steps would be taken to correct the problem. Future meetings with the users in the Consumer Banking and Corporate Banking Divisions became very hostile. Jim soon discovered that many individuals in these divisions did not see the need for the corporate database. They resented spending their time in meetings documenting the CDB requirements. They were afraid that the CDB project would lead to a shift of many of their responsibilities and functions to the Corporate Marketing Division.

Mr. Hart was also unhappy. The CDB team was spending more time than was budgeted in documenting user requirements. If this trend continued, a revised budget would have to be submitted to the Priorities Committee for approval. He was also growing tired of ordering individuals in the user departments to keep appointments with the CDB team. Mr. Hart could not understand the resistance to his project.

Jim Gunn kept trying to obtain analysts for his project but was told by his manager that none were available. Jim explained that the quality of work done by the junior analysts was not "up to par" because of lack of experience. Jim complained that he could not adequately supervise the work quality because he was forced to complete many of the analysis tasks himself. He also noted that the quality review of the user requirements subphase was scheduled for next month, making it extremely critical that experienced analysts be assigned to the project. No new personnel were assigned to the project. Jim thought about contacting the general manager again to explain his need for more experienced analysts, but did not. He was due for a semiyearly evaluation from his manager in two weeks.

Even though he knew the quality of the work was below standards, Jim was determined to get the project done on schedule with the resources available to him. He drove both himself and the team very hard during the next few weeks. The quality review of the user requirement subphase was held on schedule. Over 90 percent of the assigned tasks had to be redone before the Quality Review Board would sign-off on the review. Jim Gunn was removed as project manager.

Three senior analysts and a new project manager were assigned to the CDB project. The project received additional funding from the Priorities Committee. The user requirements subphase was completely redone despite vigorous protests from the Consumer Banking and Corporate Banking divisions.

Within the next three months the following events happened:

- The new project manager resigned to accept a position with another firm.
- John Hart took early retirement.
- The CDB project was tabled.

Synopsis

All projects at First National Bank (FNB) have project managers assigned and are handled through the Information Services Division (ISD). The organizational structure is not a matrix, although some people think that it is. The case describes one particular project, the development of a corporate database, and the resulting failure. The problem at hand is to investigate why the project failed.

Questions

1. What are the strengths of FNB?
2. What are the major weaknesses?
3. What is the major problem mentioned above? Defend your answer.
4. How many people did the project manager have to report to?
5. Did the PM remain within vertical structure of the organization?
6. Is there anything wrong if a PM is a previous co-worker of some team members before the team is formed?
7. Who made up the project team?
8. Was there any resistance to the project by company management?
9. Was there an unnecessary duplication of work?
10. Was there an increased resistance to change?
11. Was the communication process slow or fast?
12. Was there an increased amount of paperwork?
13. What are reasonable recommendations?
14. Does the company have any type of project management methodology?
15. Could the existence of a methodology have alleviated any of the above problems?
16. Did the bank perform strategic planning for project management or did it simply rush into the project?
17. Why do organizations rush into project management without first performing strategic planning for project management or, at least, some form of benchmarking against other organizations?

Index